COMPUTATION OF DIFFERENTIAL EQUATIONS AND DYNAMICAL SYSTEMS

SERIES ON APPLIED MATHEMATICS

Editor-in-Chief: Frank Hwang
Associate Editors-in-Chief: Zhong-ci Shi and Kunio Tanabe

Vol. 1 International Conference on Scientific Computation
eds. T. Chan and Z.-C. Shi

Vol. 2 Network Optimization Problems — Algorithms, Applications and Complexity
eds. D.-Z. Du and P. M. Pandalos

Vol. 3 Combinatorial Group Testing and Its Applications
by D.-Z. Du and F. K. Hwang

eries on
pplied Mathematics
olume 4

Proceedings of the
International Conference
on

COMPUTATION OF DIFFERENTIAL EQUATIONS AND DYNAMICAL SYSTEMS

Beijing, P. R. China September 1 – 5, 1992

Editors

Kang Feng & Zhong-Ci Shi
Computing Center, Academia Sinica
China

World Scientific
Singapore • New Jersey • London • Hong Kong

Published by

World Scientific Publishing Co Pte Ltd
5 Toh Tuck Link, Singapore 596224
USA office: 27 Warren Street, Suite 401-402, Hackensack, NJ 07601
UK office: 57 Shelton Street, Covent Garden, London WC2H 9HE

British Library Cataloguing-in-Publication Data
A catalogue record for this book is available from the British Library.

COMPUTATION OF DIFFERENTIAL EQUATIONS AND DYNAMICAL SYSTEMS

ISBN-13 978-981-02-1339-8
ISBN-10 981-02-1339-5

SCIENTIFIC COMMITTEE

ORGANIZING COMMITTEE

PREFACE

The International Conference on Computation of Differential Equations and Dynamical Systems which was held on September 1–5, 1992 in Beijing was sponsored by the Chinese Academy of Sciences and The National Natural Science Foundation of China. About 80 researchers attended the conference, 30 of which were invited speakers from China, US, Sweden, Germany, Japan, Italy, Spain Switzerland, Russia, Korea and Hong Kong.

This volume contains the plenary and sessional invited lectures and selected contributed papers presented at the Conference. A wide spectrum of topics on current trends in the field of computation of differential equations and dynamical systems, including various original contributions in numerical methods, analysis as well as applications to other branches of science and technology are covered.

We would like to extend our sincere thanks to all the participants of the conference for their outstanding lectures, active support and friendly cooperations which made this conference a success.

Finally, we sincerely thank World Scientific Publishing Company, Singapore for the publication of this proceedings as a volume of the Series on Applied Mathematics.

Kang FENG
Member, Academia Sinica
Chairman, Scientific Committee

Zhong-ci SHI
Member, Academia Sinica,
Chairman, Organizing Committee

Beijing, September 1992

CONTENTS

Contributed Papers

Formal Dynamical Systems and Numerical Algorithms*

FENG Kang

Academia Sinica Computing Center, Beijing 100080, China

1. Preliminaries

We study vector fields, their associated dynamical systems, phase flows together with their algorithmic approximations in $\mathbf{R}^N$ from the formal power series approach. Our considerations will be local in both space and time, all related objects are C^∞ smooth. We use coordinate description and matrix notation, the coordinate vectors in $\mathbf{R}^N$ and vector functions $a : \mathbf{R}^N \to \mathbf{R}^N$ are denoted by column matrices. The identity vector function 1_N is given by $1_N(x) = x$. For vector function $a = (a_1, \cdots, a_N)^T : \mathbf{R}^N \to \mathbf{R}^N$,

$$
\begin{aligned}
a_* &:= \left(\frac{\partial a_i}{\partial x_j}\right) = \text{Jacobian matrix of } a \\
a^* &:= \sum a_i \frac{\partial}{\partial x_i} = \text{linear differential operator of first order associated to } a.
\end{aligned}
$$

The association $a \to a^*$ is linear, a^* operates on scalar functions $\phi : \mathbf{R}^N \to \mathbf{R}$ and on vector functions $b : \mathbf{R}^N \to \mathbf{R}^N$ as

$$
\begin{aligned}
a^*\phi &= \sum a_i \frac{\partial \phi}{\partial x_i} \\
a^*b &= a^*(b_1, \cdots, b_N)^T = (a^*b_1, \cdots, a^*b_N)^T = b_*a, \quad a^*1_N = a.
\end{aligned}
$$

Multiple application of linear differential operators are naturally defined such as a^*b^*, $(a^*b^*)c^*$, $a^*(b^*c^*)$, etc. The operations are multilinear, associative but non-commutative; thus powers can be defined

$$
a^{*k} = a^*a^* \cdots a^* (k \text{ times}), \qquad a^k := a^{*k}1_N
$$

The identity operator I operates on scalar and vector functions ϕ and b as $I\phi = \phi$, $Ib = b$.

*Supported by National Natural Science Foundation of China and State Commission of Science and Technology.

We identify all vector functions $a : \mathbf{R}^N \to \mathbf{R}^N$ as vector fields. All vector fields in $\mathbf{R}^N$ form a (∞-dimensional) real Lie algebra $\mathbf{V}^N$ under Lie bracket

$$[a, b] := a_* b - b_* a = b^* a - a^* b = (b^* a^* - a^* b^*) 1_N.$$

The Lie algebra $\mathbf{V}^N$ is associated to the (∞-dimensional) local Lie group $\mathbf{D}^N$ of near-identity diffeomorphisms—or simply near-1 maps—of $\mathbf{R}^N$.

Consider the dynamical system in $\mathbf{R}^N$

$$\frac{dx}{dt} = a(x) \tag{1}$$

defined by a vector field a. It possesses a phase flow $e_a^t = e^t$, which is a one-parameter (in t) group of near-1 maps of $\mathbf{R}^N$,

$$e^0 = 1_N, \qquad e^{t+s} = e^t \circ e^s,$$

and generates the solution by $x(0) \to e_a^t x(0) = x(t)$. The phase flow is expressible as a convergent power series in t,

$$e_a^t = 1_N + \sum_{k=1}^{\infty} t^k e_k,$$

$$e_0 = 1_N, \qquad e_k = \frac{1}{k} a^* e_{k-1} = \frac{1}{k!}(a^*)^k 1_N = \frac{1}{k!} a^k.$$

We define

$$\operatorname{Exp} ta^* := I + \sum_{k=1}^{\infty} \frac{1}{k!}(ta)^{*k}, \quad I \text{ is the identity operator,} \tag{2}$$

This is an operator power series operating on scalar functions and vector functions, and define

$$\exp ta := (\operatorname{Exp} ta^*) 1_N = 1_N + \sum_{k=1}^{\infty} \frac{1}{k!}(ta^*)^k 1_N = 1_N + \sum_{k=1}^{\infty} \frac{t^k}{k!} a^k. \tag{2$'$}$$

then

$$\begin{cases} e_a^t = (\operatorname{Exp} ta^*) 1_N = \exp ta, \\ \phi \circ e_a^t = \phi \circ \exp ta = (\operatorname{Exp} ta^*)\phi, \quad \forall \text{ scaler function } \phi, \\ b \circ e_a^t = b \circ \exp ta = (\operatorname{Exp} ta^*) b = (\operatorname{Exp} ta^*) b^* 1_N, \quad \forall \text{ vector function } b. \end{cases} \tag{3}$$

Each numerical algorithm solving the system (1) possesses the *step transition map* f_a^s which is one-parameter (in step-size s) family (in general not a one-parameter group in s) of near-1 maps on R^N, expressible as a convergent power series in s

$$f_a^s = 1_N + \sum_{1}^{\infty} s^k f_k, \tag{4}$$

the coefficients can be determined recursively from the defining difference equation. The transition generates the numerical solution $x(0) \to (f_a^s)^N x(0) \approx x(Ns)$ by iterations with step-size s chosen fixed in general.

The main problem is to construct and analyse the algorithmic approximations $f_a^s \approx e_a^t|_{t=s} = e_a^s$ in a *proper* way. For this purpose we propose a *unified* framework based on the apparatus of formal power series, Lie algebra of vector fields and the corresponding Lie group of diffeomorphisms [1,2].

2. Near-0 and Near-1 formal Power Series

Among the formal power series $\sum_0^\infty s^k a_k$, $a_k : \mathbf{R}^N \to \mathbf{R}^N$, we pick out two special classes. The first class consists of those with $a_0 = 0$, called near-0 formal vector fields; the second class consists of those with $a_0 = 1_N$, called near-1 formal maps (diffeomorphisms).

All near-0 formal vector fields $a^s = \sum_1^\infty s^k a_k$ form a (∞-dim.) real Lie algebra $\mathbf{FV}^N$ under the Lie bracket

$$[a^s, b^s] = \Big[\sum_1^\infty s^k a_k, \sum_1^\infty s^k b_k\Big] := \sum_{k=2}^\infty s^k \sum_{i+j=k} [a_i, b_j].$$

The associated near-0 formal differential operators and their products are

$$(a^s)_* := \Big(\sum_1^\infty s^k a_k\Big)_* := \sum_1^\infty s^k a_{k*}$$

$$a^{s*} = \Big(\sum_1^\infty s^k a_k\Big)^* := \sum_1^\infty s^k a_k^*,$$

$$a^{s*} b^{s*} := \sum_2^\infty s^k \sum_{i+j=k} a_i^* b_j^*, \quad (a^{s*})^2 := a^{s*} a^{s*}, \text{ etc.}$$

For any vector function $a = (a_1, \cdots, a_N)^T : \mathbf{R}^N \to \mathbf{R}^N$ and any near-1 formal map $g^s = 1 + \sum s^k g_k$ we define the composition

$$(a \circ g^s)(x) = a(g^s(x)) = a(x) + \sum_1^\infty s^k (a \circ g)_k(x),$$

$$(a \circ g)_k = \sum_{m=1}^k \sum_{k_1 + \cdots + k_m = k} \frac{1}{m!} (D^m a)(g_{k_1}, \cdots, g_{k_m}), \tag{5}$$

where

$$D^m a = (D^m a_1, \cdots, D^m a_N)^T,$$

$$D^m a_i(v_1, \cdots, v_m) = \sum_{j_1, \cdots, j_m = 1}^N \frac{\partial^m a_i}{\partial x_{j_1} \cdots \partial x_{j_m}} v_{1j_1} \cdots v_{mj_m}$$

is the usual m-th differential multi-linear form for m tangent vectors $v_i = (v_{i_1}, \cdots, v_{i_N})^T$, $i = 1, \cdots, m$ at point $x \in \mathbf{R}^N$ which is invariant under permutation of vectors. Using the identities

$$\begin{aligned}
(D^1 a)(b) &= b^* a, \\
(D^2 a)(b, c) &= (c^* b^* - (c^* b)^*) a, \\
(D^3 a)(b, b, b) &= (b^{*3} + 2b^{3*} - 3b^* b^{2*}) a,
\end{aligned}$$

we get in particular

$$\begin{aligned}
(a \circ g)_1 &= g_1^* a, \\
(a \circ g)_2 &= g_2^* a + \frac{1}{2}(g_1^{*2} - g_1^{2*}) a, \\
(a \circ g)_3 &= g_3^* a + ((g_2^* g_1^* - (g_2^* g_1)^*) a + \frac{1}{3!}(g_1^{*3} + 2g_1^{3*} - 3g_1^* g_1^{2*}) a.
\end{aligned}$$

For any two near-1 formal maps $f^s = 1_N + \sum s^k a_k$, $g^s = 1 + \sum s^k g_k$, the composition $f^s \circ g^s$ is defined in a term by term way:

$$\begin{aligned}
(f^s \circ g^s)(x) &= f^s(g^s(x)) = 1_N(g^s(x)) + \sum_{k=1}^{\infty} s^k f_k(g^s(x)) \\
&=: 1_N(x) + \sum_{k=1}^{\infty} s^k (f \circ g)_k(x), \\
(f \circ g)_1 &= f_1 + g_1, \qquad (6) \\
(f \circ g)_k &= f_k + g_k + \delta(f_1, \cdots, f_{k-1}; g_1, \cdots, g_k), \quad k \geq 2, \\
\delta(f_1, \cdots, f_{k-1}; g_1, \cdots, g_{k-1}) &= \sum_{i=1}^{k-1} \sum_{m=1}^{i} \sum_{i_1 + \cdots + i_m = i} \frac{1}{m!}(D^m f_{k-i})(g_{i_1}, \cdots, g_{i_m}),
\end{aligned}$$

we get in particular,

$$\begin{aligned}
(f \circ g)_2 &= f_2 + g_2 + g_1^* f_1, \\
(f \circ g)_3 &= f_3 + g_3 + g_1^* f_2 + g_2^* f_1 + \frac{1}{2}(g_1^{*2} - g_1^{2*}) f_1, \\
(f \circ g)_4 &= f_4 + g_4 + g_1^* f_3 + g_2^* f_2 + g_3^* f_1 + \frac{1}{2}(g_1^{*2} - g_1^{2*}) f_2 \\
&\quad + (g_2^* g_1^* - (g_2^* g_1)^*) f_1 + \frac{1}{3!}(g_1^{*3} + 2g_1^{3*} - 3g_1^* g_1^{2*}) f_1.
\end{aligned}$$

Under this composition rule, all near-1 formal maps $f^s = 1 + \sum_1^\infty s^k f_k$ form (∞-dim) formal Lie group $\mathbf{FD}^N$. In group $\mathbf{FD}^N$ inverse elements, square roots, rational powers, etc., always exist, their coefficients can always be determined recursively by the defining composition relations. For example, the inverse $(f^s)^{-1} := 1 + \sum s^k h_k = h^s$ is defined by $(f^s \circ h^s) = 1_N$, hence

$$f_1 + h_1 = 0, \quad f_k + h_k + \delta(f_1, \cdots, f_{k-1}; h_1, \cdots, h_{k-1}) = 0, \quad k \geq 2.$$

In particular,

$$h_1 = -f_1, \quad h_2 = -f_2 + f_1^2, \quad h_3 = -f_3 + f_1^* f_2 + (f_2^* - f_1^{2*}) f_1 - \frac{1}{2} f_1^3 + \frac{1}{2} f_1^{2*} f_1.$$

There is an obvious one-one correspondence between the Lie algebra $\mathbf{FV}^N$ and the Lie group $\mathbf{FD}^N$, established simply by $+1_N$ and by -1_N. However, the more significant one-one correspondence between them is given by exp and its inverse log.

$$\begin{aligned}
&\exp : \mathbf{FV}^N \to \mathbf{FD}^N \\
&a^s = \sum_1^\infty s^k a_k \to \exp a^s := 1_N + \sum_{m=1}^\infty \frac{1}{m!} (a^{s*})^m 1_N =: 1_N + \sum_1^\infty s^k f_k = f^s. \quad (7)
\end{aligned}$$

Note that

$$(a^{s*})^m = \left(\sum s^{k_1} a_{k_1}^*\right) \cdots \left(\sum s^{k_m} a_{k_m}^*\right) = \sum_{k_1, \cdots, k_m = 1}^\infty s^{k_1 + \cdots + k_m} a_{k_1}^* \cdots a_{k_m}^*,$$

so we get easily

$$\begin{aligned}
f_k &= \sum_{m=1}^k \frac{1}{m!} \sum_{k_1 + \cdots + k_m = k} a_{k_1}^* \cdots a_{k_m}^* 1_N, \qquad k \geq 1, \quad f_1 = a_1, \qquad (8) \\
f_k &= a_k + \sum_{m=2}^k \frac{1}{m!} \sum_{k_1 + \cdots + k_m = k} a_{k_1}^* \cdots a_{k_m}^* 1_N, \qquad k \geq 2, \quad f_2 = a_2 + \frac{1}{2} a_1^2.
\end{aligned}$$

Note that (8) provides a 2-way recursion formula from $a_1, \cdots, a_k$ to $f_1, \cdots, f_k$ and vice versa. Therefore exp maps $\mathbf{FV}^N$ one-one onto $\mathbf{FD}^N$ and its inverse, i.e., log is defined by the same (8):

$$\log = (\exp)^{-1} : \mathbf{FD}^N \to \mathbf{FV}^N, \quad \log \exp a^s = a^s, \quad \exp \log f^s = f^s.$$

In particular

$$\begin{aligned}
a_1 &= f_1, \quad a_2 = f_2 - \frac{1}{2} a_1^2, \quad a_3 = f_3 - \frac{1}{2}(a_1^* a_2 + a_2^* a_1) - \frac{1}{3!} a_1^3, \\
a_4 &= f_4 - \frac{1}{2}(a_1^* a_3 + a_2^2 + a_3^* a_1) - \frac{1}{3!}(a_1^* a_1^* a_2 + a_1^* a_2^* a_1 + a_2^* a_1^* a_1) - \frac{1}{4!} a_1^4, \qquad (9) \\
a_k &= f_k - \sum_{m=2}^{k-1} \frac{1}{m!} \sum_{k_1 + \cdots + k_m = k} a_{k_1}^* \cdots a_{k_m}^* 1_N - \frac{1}{k!} a_1^k, \quad k \geq 3.
\end{aligned}$$

An equivalent construction of $\log f^s = a^s$ is

$$\log f^s = \sum_{m=1}^\infty \frac{(-1)^{m-1}}{m} h_m^s, \quad \text{where } h_1^s = f^s - 1_N, \quad h_m^s = h_{m-1}^s \circ f^s - h_{m-1}^s. \quad (10)$$

It is easy to compute

$$\begin{aligned} h_1^s &= \sum_{k=1}^{\infty} s^k f_k = \sum_{k_1=1}^{\infty} s^{k_1}(1_N \circ f)_{k_1} \\ h_2^s &= \sum_{k_1=k_2=1}^{\infty} s^{k_1+k_2}((1_N \circ f)_{k_1} \circ f)_{k_2}, \\ h_3^s &= \sum_{k_1,k_2,k_3=1}^{\infty} s^{k_1+k_2+k_3}(((1_N \circ f)_{k_1} \circ f)_{k_2} \circ f)_{k_3}, \text{ etc.} \end{aligned}$$

Substituting in (10) and equating with $\sum_1^{\infty} s^k a_k$, we get

$$a_k = \sum_{m=1}^{k} \frac{(-1)^{m-1}}{m} \sum_{k_1+\cdots+k_m=k} (\cdots((1_N \circ f)_{k_1} \circ f)_{k_2} \cdots \circ f)_{k_m}. \tag{11}$$

It is easy to verify $\log \exp a^s = a^s$ for this log, so this is precisely the inverse of exp, thus agrees with the previous one.

We use the above construction (11) to establish the *formal* Campbell-Hausdorff formula:

For arbitrary near-1 formal maps f^s, g^s

$$\log(f^s \circ g^s) = \log f^s + \log g^s + \sum_{k=1}^{\infty} d_k(\log f^s, \log g^s), \tag{12}$$

where, for $\log f^s = a^s$, $\log g^s = b^s$,

$$d_k(a^s, b^s) = \frac{1}{k} \sum_{m=1}^{k} \frac{(-1)^{m-1}}{m} \sum_{\substack{p_1+q_1+\cdots+p_m+q_m=k \\ p_i+q_i \geq 1, p_i \geq 0, q_i \geq 0}} \frac{[(a^s)^{p_1}(b^s)^{q_1} \cdots (a^s)^{p_m}(b^s)^{q_m}]}{p_1! q_1! \cdots p_m! q_m!} \tag{13}$$

where

$$(x)^p = xx\cdots x \ (p\text{-times}), \quad [x_1 x_2 x_3 \cdots x_n] = [[\cdots[[x_1, x_2], x_3], \cdots], x_n].$$

In particular

$$d_1 = \frac{1}{2}[a^s, b^s], \quad d_2 = \frac{1}{12}([a^s b^s b^s] + [b^s a^s a^s]), \quad d_3 = -\frac{1}{24}[a^s b^s b^s a^s].$$

Let $\log(f^s \circ g^s) = c^s = \sum_1^{\infty} s^k c_k$. Then

$$\begin{aligned} c_1 &= a_1 + b_1, \qquad c_2 = a_2 + b_2 + \frac{1}{2}[a_1 b_1], \\ c_3 &= a_3 + b_3 + \frac{1}{2}([a_1 b_2] + [a_2 b_1]) + \frac{1}{12}([a_1 b_1 b_1] + [b_1 a_1 a_1]), \end{aligned}$$

$$\begin{aligned} c_4 &= a_4 + b_4 + \frac{1}{12}([a_1b_3] + [a_2b_2] + [a_3b_1]) \\ &+ \frac{1}{12}([a_1b_1b_2] + [a_1b_2b_1] + [a_2b_1b_1] + [b_1a_1a_2] + [b_1a_2a_1] + [b_2a_1a_1]) \\ &- \frac{1}{24}[a_1b_1b_1a_1], \quad \text{etc.} \end{aligned}$$

Note that the classical CH formula is restricted to the composition of two one-parameter groups, where $\log f^s = sa_1$, $\log g^s = sb_1$.

The log transform reduces matters at the Lie group level to those at the easier level of Lie algebra. All properties of near-1 formal maps have their logarithmic interpretations. We list some of them, let $\log f^s = a^s = \sum_1^\infty s^k a_k$:

1. f^s is a phase flow, i.e., $f^{s+t} = f^s \circ f^t \iff \log f^s = sa_1$.
2. f^s is revertible, i.e., $f^s \circ f^{-s} = 1_N \iff \log f^s$ is odd in s.
3. f^s raised to real μth power $(f^s)^\mu \iff \log(f^s)^\mu = \mu \log f^s$. In particular $\log(f^s)^{-1} = -\log f^s$, $\log \sqrt{f^s} = \frac{1}{2}\log f^s$.
4. f^s scaled to $f^{\alpha s} \iff \log(f^{\alpha s}) = (\log f)^{\alpha s}$. In particular $\log(f^{-s}) = (\log f)^{-s}$.
5. $f^s - g^s = O(s^{p+1}) \iff \log f^s - \log g^s = O(s^{p+1})$.
6. $f^s \circ g^s = g^s \circ f^s \iff [\log f^s, \log g^s] = 0 \iff \log(f^s \circ g^s) = \log f^s + \log g^s$.
7. $(f^s \circ g^s) = h^s \iff \log h^s = \log(f^s \circ g^s) = \log f^s + \log g^s + \sum_1^\infty d_k(\log f^s, \log g^s)$.
8. f^s symplectic $\iff$ all a_k are Hamiltonian fields.
9. f^s contact $\iff$ all a_k are contact fields.
10. f^s volume-preserving $\iff$ all a_k are source-free fields.

The log transform has important bearing on dynamical systems with Lie algebra structure, the structure-preserving property of maps f^s at the Lie group $(\mathbf{G} \subset \mathbf{D}_m)$ level can be characterized through their logarithms at the associated Lie algebra $(\mathbf{L} \subset \mathbf{V}_m)$ level, see [2, 6].

3. Algorithmic approximations to phase flows

We return to the main problem of approximation to the phase flow for dynamical system $dx/dt = a(x)$.

$$f_a^s = f^s = 1_N + \sum s^k f_k \approx e_a^s = 1_N + \sum s^k e_k, \quad e_k = a^k/k!. \tag{14}$$

If $f_k = e_k$, $1 \le k \le p$, we say f_a^s is accurate to order $\ge p$, if moreover, $f_{p+1} \ne e_{p+1}$, we say it is accurate to order p.

Let $\log f^s = a^s = \sum s^k a_k$. Note that the first $p+1$ equations in (8) completely determine $a_1, a_2, \cdots, a_{p+1}$ and $f_1, f_2, \cdots, f_{p+1}$ each other. It is then easy to establish

$$\begin{aligned} &f_k = e_k, \quad 1 \le k \le p; \quad f_{p+1} \ne e_{p+1} \iff \\ &a = a_1 = e_1; \quad a_k = 0, \quad 1 < k \le p; \quad a_{p+1} = f_{p+1} - e_{p+1} \ne 0. \end{aligned} \tag{15}$$

So the orders of approximation for $f_a^s \approx e_a^s$ and for $\log f_a^s - sa$ are the same.

Moreover, note that we have a *formal field*

$$s^{-1}\log f^s = s^{-1}a^s = a + \sum_1^\infty s^k a_{k+1} = a + O(s^p) \tag{16}$$

which is the original field a up to a near-0 perturbation and defines a *formal dynamical system*

$$\frac{dx}{dt} = (s^{-1}\log f^s)(x) = a(x) + \sum_1^\infty s^k a_{k+1}(x) \tag{17}$$

having a *formal phase flow* (in two parameters t and s with group property in t) $e^t_{s^{-1}a^s} = \exp ts^{-1}a^s$ whose *diagonal formal flow* $e^t_{s^{-1}a^s}|_{t=s}$ is exactly f^s. This means that any compatible algorithm f_a^s of order p gives perturbed solution of a right equation with field a; however, it gives the right solution of a perturbed equation with field $s^{-1}\log f_a^s = a + O(s^p)$. There could be many methods with the same formal order of accuracy but with quite different qualitative behavior. The problem is to choose among them those leading to allowable perturbations in the equation. For systems with geometric structure, the propositions 8, 9 and 10 provide guidelines for a proper choice. The structure-preservation requirement for the algorithms precludes all unallowable perturbations alien to the pertinent type of dynamics. Take, for example, Hamiltonian systems. A transition map f_a^s for Hamiltonian field a is symplectic if and only if all fields a_k are Hamiltonian, i.e., the induced perturbations in the equation are Hamiltonian [2,6]. So symplectic algorithms are clean, inherently free from all kinds of perturbations alien to Hamiltonian dynamics (such as artificial dissipation inherent in the vast majority of conventional methods), this accounts for their superior performance. The situations are the same for contact and volume-preserving algorithms [3–6].

Finally we give, as an illustration, four simplest methods together with step transition maps and their logarithms.

$$e_a^s = 1_N + sa + \frac{1}{2}s^2a^2 + \frac{1}{3!}s^3a^3 + O(s^4).$$

1. Explicit Euler method E

$$\begin{aligned} x_1 - x_0 &= sa(x_0), \\ f^s - 1_N &= sa, \\ f_E^s &= 1_N + sa, \\ \log f_E^s &= sa - \frac{s^2}{2}a^2 + O(s^3), \end{aligned}$$

non-revertible, ord $= 1$.

2. Implicit Euler method I

$$
\begin{aligned}
&x_1 - x_0 = sa(x_1),\\
&f^s - 1_N = sa \circ f^s,\\
&f_I^s = (1_N - sa)^{-1} = (f_E^{-s})^{-1} = sa + s^2a^2 + O(s^3),\\
&\log f_I^s = sa + \frac{s^2}{2}a^2 + O(s^3),
\end{aligned}
$$

non-revertible, ord $= 1$.

3. Trapezoidal method T

$$
\begin{aligned}
x_1 - x_0 &= \frac{s}{2}(a(x_1) + a(x_0)),\\
f^s - 1_N &= \frac{s}{2}(a \circ f^s + a),\\
f_T^s &= \left(1_N - \frac{s}{2}a\right)^{-1} \circ \left(1_N + \frac{s}{2}a\right) = f_I^{s/2} \circ f_E^{s/2}\\
&= (f_E^{s/2})^{-1} \circ f_C^s \circ f_E^{s/2} = 1_N + sa + \frac{s^2}{2}a^2 + \frac{s^3}{4}a^3 + O(s^4),\\
\log f_T^s &= sa + \frac{s^3}{12}a^3 + O(s^5),
\end{aligned}
$$

revertible, ord $= 2$, symplectic for linear Hamiltonian but non-symplectic for non-linear Hamiltonian systems.

4. Centered Euler method C

$$
\begin{aligned}
&x_1 - x_0 = sa\Big(\frac{1}{2}(x_1 + x_0)\Big),\\
&f^s - 1_N = sa \circ \Big(\frac{1}{2}(f^s + 1_N)\Big),
\end{aligned}
$$

2-stage version recommended for implementation

$$
\begin{aligned}
\bar{x} &= x + \frac{s}{2}a(\bar{x}), \qquad x_1 = 2\bar{x} - x_0,\\
\bar{x} &= f_I^{s/2}(x_0), \quad x_1 = 2f_I^{s/2}(x_0) - 1_N(x_0),\\
f_C^s &= 2f_I^{s/2} - 1_N = \left(1_N + \frac{s}{2}a\right)^{-1} \circ \left(1_N - \frac{s}{2}a\right) = f_E^{s/2} \circ f_I^{s/2}\\
&= 1_N + sa + \frac{s^2}{2}a^2 + \frac{s^3}{8}(a_*a^2 + a^3) + O(s^4),\\
\log f_C^s &= sa + s^3\Big(\frac{1}{8}a_*a^2 - \frac{1}{24}a^3\Big) + O(s^5),
\end{aligned}
$$

revertible, ord $= 2$, unconditionally symplectic with preservation of all quadratic invariants for Hamiltonian systems.

Note the similarities and delicate differences between C and T: Both can be composed by a $s/2$ implicit and a $s/2$ explicit stages but in opposite orderings.

Moreover, they are conjugate to each other, noted earlier independently in [7]. C is far less known than T, it becomes prominent only after the recent development of symplectic algorithms [3]. In crucial aspects C is superior.

Acknowledgement. The author is grateful to Dr. Wang Dao-liu for discussions and valuable help in preparing the manuscript.

References

[1] Feng Kang, The calculus of formal power series for diffeomorphisms and vector fields, Preprint, Acad. Sinica Comp. Ctr., 1991.

[2] Feng Kang, Formal power series and numerical algorithms for dynamical systems, *Proc. Inter. Conf. Scientific Computation*, Hangzhou, 1991, Ed. T. Chan, Z.C. Shi, World Scientific, Singapore, 1992, 28–35.

[3] Feng Kang, On difference schemes and symplectic geometry, *Proc. 1984 Beijing Symp. on Diff. Geometry and Diff. Equations*, Ed. Feng Kang, Science Press, Beijing, 42–58.

[4] Feng Kang, The Hamiltonian way for computing Hamiltonian dynamics, in *Applied and Industrial Mathematics, Proc. Venice Symp.*, 1990, 17–35, ed. R. Spigler, Kluwer, the Netherlands, 1991.

[5] Feng Kang, Qin Meng Zhao, *Comput. Phys. Comm.*, **65**(1991), 173–187.

[6] Feng Kang, Symplectic, contact and volume-preserving algorithms, *Proc. 1st China-Japan Conference on Numerical Mathematics*, Beijing, 1992, World Scientific, Singapore, 1993.

[7] Qin Meng Zhao, Oral communications.

Finite Volume Methods for Conservation Laws

Dietmar Kröner
Institut für Angewandte Mathematik, Universität Freiburg
Hermann-Herder-Str. 10, D-7800 Freiburg

Abstract

In this paper we shall report on some new results concerning the convergence and numerical experiments for finite volume methods on unstructured grids for solving equations and systems of conservation laws in several space dimensions.

1 Introduction

In this paper we shall give a survey of recent results concerning finite volume discretizations for hyperbolic conservation laws in two space dimensions. We consider

$$\partial_t u + \partial_x f_1(u) + \partial_y f_2(u) = 0 \quad \text{in} \quad \mathbf{R}^2 \times \mathbf{R}^+ \tag{1}$$

with initial values

$$u(x, y, 0) = u_0(x, y) \quad \text{in} \quad \mathbf{R}^2 \tag{2}$$

where $u : \mathbf{R}^2 \times \mathbf{R}^+ \to \mathbf{R}^\mathrm{m}$, $f_1, f_2 : \mathbf{R}^\mathrm{m} \to \mathbf{R}^\mathrm{m}$ and $u_0 \in L^\infty(\mathbf{R}^2, \mathbf{R}^\mathrm{m})$. Essentially there are three different numerical methods which are used in general for solving (1), (2). There are the finite difference methods on cartesian grids in the form of dimensional splitting (or fractional steps) [3], [4], the finite volume [2], [11] and the finite element methods [9], [18] on arbitrary unstructured meshes. The last two schemes are very appropriate for selfadapting grid refinement and they can be fit to general geometries very easily . For finite difference methods Crandall and Majda [4] have proved convergence results for monotone dimensional splitting methods in several space dimensions. Johnson and Szepessy [9] have used the method of measure valued solutions in the sense of DiPerna [5] [18] for the convergence proof of the streamline diffusion shock capturing method. Coquel and LeFloch have applied the same method in [3] for special dimensional splitting schemes of higher order of accuracy. For the proof in [3], the authors analyse the local entropy dissipation in the single elementary waves of the Lax-Friedrichs and the Godunov scheme in order to get an estimate which is weaker then the BV-estimate. In a more recent paper [2] they could also generalize these ideas to finite volume schemes under some special assumptions. They need a very restrictive condition either for the triangulation or for the numerical scheme. In the latter case they have to assume that the numerical viscosity is estimated from below. In [13] convergence was shown for a more general class of finite volume methods. In this paper it is not necessary to assume the strong conditions as in [2]. Also the control of the entropy dissipation in [13] is completely different to that in [2].

In this paper we shall give an overview of the results in [1], [7],[12],[13] where the reader can also find the details of the proofs. First let us describe the conditions which we need for the triangulation and some notations.

DEFINITION 1.1 *Let a k-polygon be a closed, convex polygon with k vertices. The set* $T := \{T_i \mid T_i \text{ is a k-polygon for } i \in I \subseteq \mathbf{N}\}$, *(where* $I \subseteq \mathbf{N}$ *is an index set) is called an unstructured grid of* $\Omega \subset \mathbf{R^n}$ *if the following two properties are satisfied:*

$$1) \quad \Omega = \bigcup_{i \in I} T_i \,.$$

$$2) \quad \text{For two different } T_i, T_j \text{ we have } T_j \cap T_i = \emptyset \text{ or}$$
$$T_j \cap T_i = \text{ a common vertex of } T_i, T_j \text{ or}$$
$$T_j \cap T_i = \text{ a common edge of } T_i, T_j .$$

NOTATION 1.2 *Let* $(T_j)_j$ *denote an unstructured grid of the* $\mathbf{R^2}$. *In particular we shall use the following notation:*

T_j : *the* j^{th} *cell of the unstructured grid.*

$\mid T_j \mid$: *area of* T_j.

w_j : *center of gravity of* T_j.

$T_{jl}, l = 1, \ldots, k$: *neighboring cells of* T_j.

$\alpha(j,l)$: *global number of the* l^{th}*-neighboring cell* T_{jl} *of* T_j *such that* $T_{jl} = T_{\alpha(j,l)}$.

$w_{jl}, l = 1, \ldots, k$: *center of gravity of* T_{jl}, $w_{jl} = w_{\alpha(j,l)}$.

u_j^n : *approximation of the exact solution u on* T_j *at time* $n\Delta t$. *The function* u_j^n *is assumed to be constant on* T_j .

u_{jl}^n : *approximation of the exact solution u on* T_{jl} *at time* $n\Delta t$; $u_{jl}^n = u_{\alpha(j,l)}^n$.

$u_j := u_j^n$.

S_{jl} : l^{th}*-edge of* T_j.

z_{jl} : *midpoint of the l-th edge of the cell j.*

ν_{jl} : *outer normal to* S_{jl} *of length* $\mid S_{jl} \mid$.

$n_{jl} := \frac{\nu_{jl}}{|\nu_{jl}|}$. *By* n_{jlx} *and* n_{jly} *we denote the* $x-$ *and* $y-$ *coordinate of* n_{jl}*, respectively.*

$h := \sup_{j \in I} \text{diam}(T_j)$.

$u^n(z) := u_j^n \quad \textit{if} \quad z \in T_j$.

$u_h(z,t) := u_j^n \quad \textit{if} \quad z \in T_j \quad \textit{and} \quad t_n \leq t < t_{n+1}, \quad n = 0, 1, \ldots$

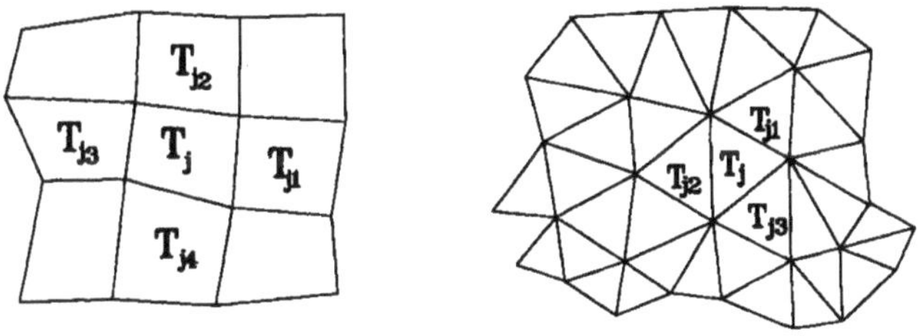

Figure 1

Figure 1 illustrates the notation for unstructured grids of quadrangles ($k = 4$) and triangles ($k = 3$), respectively. Also, the dual cells of a cell-vertex mesh can be considered as an unstructured cell-centered grid.

ASSUMPTION 1.3 *For any* j *and* l *there exists a unique number* m *such that for* $i := \alpha(j,l)$ *we have:* $\alpha(i,m) = j$*.*

We use the notation $\nu_{j,\alpha(j,l)} := \nu_{jl}$ and therefore it turns out that $\nu_{im} = -\nu_{jl}$.

ASSUMPTION 1.4 *For all* $l = 1, \ldots, k$ *we assume that for any bounded domain* Ω

$$\sum_{T_j \cap \Omega \neq \emptyset} || T_j | - | T_{jl} || = o(1) \tag{3}$$

if $h \to 0$ *and that there are constants* $c_1, c_2 \geq 0$ *such that*

$$0 < c_1 \leq \frac{\Delta t}{h} \leq c_2 \tag{4}$$

if $\Delta t, h \to 0$*. Moreover, we assume that there exists a constant* $c_V > 0$ *such that*

$$\sup_j \frac{h^2}{| T_j |} \leq c_V. \tag{5}$$

2 Finite volume schemes

First let us define the general form of a finite volume scheme. The definition and the results will be given in $\mathbf{R}^2$ but they also hold in arbitrary space dimensions. For the proofs in this part we refer to [7].

DEFINITION 2.1 *For given initial values $u_0 \in L^\infty(\mathbf{R}^2, \mathbf{R}^m)$ let u_j^n be defined by the following numerical scheme:*

$$u_j^0 := \frac{1}{| T_j |} \int_{T_j} u_0 , \quad u_j^{n+1} := u_j^n - \frac{\Delta t}{| T_j |} \sum_{l=1}^{k} g_{jl}(u_j^n, u_{jl}^n) , \tag{6}$$

where for $g_{jl}, l = 1, \ldots, k$ we assume that for any $R > 0$ and for all $u, v, u', v' \in B_R(0)$ we have

$$| g_{jl}(u,v) - g_{jl}(u',v') | \leq c(R)\, h\, (| u - u' | + | v - v' |) , \tag{7}$$

$$g_{j,\alpha(j,l)}(u,v) = -g_{\alpha(j,l),j}(v,u) , \tag{8}$$

$$g_{j,\alpha(j,l)}(u,u) = \nu_{j,\alpha(j,l)} f(u) , \tag{9}$$

where $f(u) := \begin{pmatrix} f_1(u) \\ f_2(u) \end{pmatrix}$ and where we have used the notation $g_{j,\alpha(j,l)} := g_{jl}$.

There are several examples for numerical fluxes $g_{jl}(u,v)$ for which the conditions (7), (8), (9) are satisfied. First let us consider the modification of the Engquist-Osher flux to 2-D. We assume for the first and second example that (1) is a scalar conservation law in 2–D. Let $c_{jl}(u) := n_{jl} f(u)$, and define

$$c_{jl}^+(u) := \int_0^u \max\left(c_{jl}'(s), 0\right) ds , \qquad c_{jl}^-(u) := \int_0^u \min\left(c_{jl}'(s), 0\right) ds .$$

The Engquist-Osher numerical flux is then given by ([6]):

$$g_{jl}^{EO}(u,v) := | S_{jl} | \left(c_{jl}^+(u) + c_{jl}^-(v)\right) . \tag{10}$$

Now, the Lax-Friedrichs numerical flux is given explicitly by

$$g_{jl}^{LF}(u,v) := \frac{1}{2}[\nu_{jl} f(u) + \nu_{jl} f(v)] - \frac{1}{2\lambda_{jl}}(v-u) , \tag{11}$$

where λ_{jl} are arbitrarily chosen constants satisfying

$$\lambda_{jl} = \lambda_{lj} > \tilde{c} > 0\,, \quad \lambda_{jl} \sup_u \left(\nu_{jl} f'(u)\right) \leq 1\,.$$

Then g_{jl}^{EO} g_{jl}^{LF} satisfy the conditions (7), (8), (9).

In the third example let us consider the modification of the flux-vector-splitting [16],[19] to 2-D. Assume that the nonlinearities in (1) satisfy

$$f_1(u) = u f_1'(u) \quad \text{and} \quad f_2(u) = u f_2'(u). \tag{12}$$

This for instance is valid for the Euler equation of gasdynamics (see [16]). Let $C_{jl}(w) := n_{jl} f'(w)$. Since the system (1) is hyperbolic there exists a nonsingular matrix $Q(= Q_{jl}(w))$ such that $Q^{-1} C_{jl}(w) Q = D = diag(d_1, \ldots, d_n)$ where $D = D(w)$ is a diagonal matrix consisting of the eigenvalues $d_i = d_i(w)$ of $C_{jl}(w)$. Now let $D^+ := diag(d_1^+, \ldots, d_n^+)$ and $C_{jl}^+(w) := QD^+Q^{-1}$. The matrix $C_{jl}^-(w)$ is defined similarly. Then the flux-vector splitting can be written in the form (6) with

$$g_{jl}(u,v) = \mid S_{jl} \mid \left(C_{jl}^+(w_1) u + C_{jl}^-(w_2) v \right) \tag{13}$$

where

$$w_1 = u, \quad w_2 = v \quad \text{for the scheme of Steger and Warming [16],} \tag{14}$$

$$w_1 = w_2 = \frac{u+v}{2}, \quad \text{for the scheme of Vijayasundaram [19].} \tag{15}$$

The schemes defined by (13), (14) and (13), (15), respectively satisfy the conditions (7), (8), (9).

Now we have to justify the definition (6) of finite volume schemes. This means that we have to show that (6) is a discretization of the conservation law (1). Sometimes this is done by analyzing the local truncation error. But for upwind discretization on unstructured meshes in general this makes no sense since local truncation error estimates cannot be expected. In particular we have the following result. Let $(T_j)_j$ be an unstructured grid satisfying 1.2, 1.3, and 1.4, let $g_{jl}(u,v)$ be the numerical flux for the numerical scheme (6), satisfying (7), (8), (9). Then in general the scheme (6) is not pointwise consistent. This means that if u is a smooth solution of (1), $u^n := u(., t_n)$ need not necessarily satisfy

$$\frac{u_j^{n+1} - u_j^n}{\Delta t} + \frac{1}{\mid T_j \mid} \sum_{l=1}^{k} g_{jl}(u_j^n, u_{jl}^n) = O(h). \tag{16}$$

A counterexample (see [7]) which proves this statement can be given for the Engquist-Osher scheme for the linear convection equation

$$\partial_t u + a\partial_x u + b\partial_y u = 0 \quad \text{in} \quad \mathbf{R}^2 \times \mathbf{R}^+. \tag{17}$$

For the triangulation we use a mesh consisting of equalsided triangles. In the discretization (6) there are formally three contributions to the update of the new timestep in a single triangle. But if there is a special constellation between the direction (a,b) of the flux and the normals of the considered triangle it remains only one term in the sum of (6) and that is not enough to get a consistent discretization of the flux terms in (17). Therefore we have to look for a better justification for the discretization (6). Of course the best one would be to prove convergence but for systems in general this is still a big problem. Later on we shall do this for scalar conservation laws. But first we shall present a Lax-Wendroff type theorem for weak solutions. The conservation property of the scheme (6) works like a global consistency condition. The following generalization of the classical Lax-Wendroff theorem will show that if there is a L^2-convergent sequence defined by a scheme of the form (6), then the limit will be a solution of the system (1) in the distributional sense, i.e. the following vector identity holds:

$$\int_{\mathbf{R}^2\times\mathbf{R}^+} u\partial_t\phi + \int_{\mathbf{R}^2\times\mathbf{R}^+} (f_1(u)\partial_x\phi + f_2(u)\partial_y\phi) - \int_{\mathbf{R}^2} u_0\phi(\cdot,0) = 0 \tag{18}$$

for all test functions $\phi \in C_0^\infty(\mathbf{R}^2 \times [0,\infty[)$.

THEOREM 2.2 *Let us assume that there is a mesh satisfying (1.1), (1.2), (1.3) and (1.4) and initial values $u_0 \in L^\infty(\mathbf{R}^2, \mathbf{R}^m)$. Then let u_j^n be defined by the numerical scheme of the form (6) as in Definition (2.1) and u_h be defined as in Notation 1.2. Furthermore we assume*

$$\sup_h \sup_{x,y,t} | u_h(x,y,t) | \leq const, \tag{19}$$

$$\| u - u_h \|_{L^2(\mathbf{R}^2\times\mathbf{R}^+)} \to 0. \tag{20}$$

Then u is a solution of (1) in the distributional sense .

For proving Theorem (2.2) we have to multiply (6) by $\phi(w_j, t_n)$ where ϕ is a smooth test function and to sum up with respect to n and j (see [7]).

This theorem only ensures that the limit will be a weak solution of the conservation law. In general it is not clear if this limit also satifies the entropy condition. Under additional assumption this can be obtained for scalar conservation laws. Let us now consider (1) as a scalar conservation law. We have to assume that the scheme defined in (6) is monotone. This means that the function

$$H(u_j, u_{j1}, \ldots, u_{jk}) := u_j^n - \frac{\Delta t}{|T_j|} \sum_{l=1}^{k} g_{jl}(u_j^n, u_{jl}^n) \tag{21}$$

is monotone nondecreasing in any argument $u_j, u_{j1}, \ldots, u_{jk}$. For instance the Engquist-Osher finite volume scheme as defined in (6) is monotone for small Δt. Let u_j^n be defined by (6) with g_{jl} as in example (10) and let Δt such that

$$\sup_j \frac{\Delta t}{|T_j|} \sum_{l=1}^{k} \max\{\nu_{jl} f'(u_j^n), 0\} \leq 1. \tag{22}$$

Then the scheme (6) is monotone. For monotone schemes we can show that the limit will satisfy the entropy condition.

THEOREM 2.3 *Consider (1) as a scalar conservation law. In addition to the assumption of Theorem (2.2) let us assume that the scheme (6) is monotone (see (21), (22)). Then the solution u defined as in Theorem (2.2) satisfies the Kruzkov entropy condition (see [14])*

$$\int_{\mathbf{R}^2 \times \mathbf{R}^+} |u - c| \, \partial_t \phi + \int_{\mathbf{R}^2 \times \mathbf{R}^+} sign(u - c) \, (f_1(u) - f_1(c)) \, \partial_x \phi + sign(u - c) \, (f_2(u) - f_2(c)) \, \partial_y \phi \geq 0 \tag{23}$$

for all test functions $\phi \in C_0^\infty(\mathbf{R}^2 \times \mathbf{R}^+)$, $\phi \geq 0$ *and all* $c \in \mathbf{R}$.

For the proof we refer to [7].

3 Convergence of upwind finite volume schemes

While the results of Theorem (2.2) and (2.3) are also valid for systems we have to restrict ourselves to scalar conservation laws if we want to prove convergence of the finite volume scheme (6). In this paragraph we shall describe a class of upwind finite volume schemes for which we can prove convergence. In particular, we show that the Engquist-Osher and the Lax-Friedrichs schemes with suitably chosen numerical entropy fluxes belong to that class. For the proofs in this part we refer to [13]. First we shall describe some conditions, which are sufficient for convergence.

THEOREM 3.1 *Let us assume that 1.3, (4) and (5) are satisfied. Let* $u_0 \in L^\infty(\mathbf{R}^2) \cap L^1(\mathbf{R}^2)$ *and let* u_j^n *be defined by the numerical scheme (6) such that* *for g_{jl} the properties (7),(8) and (9). We assume that for any $U \in C^2(\mathbf{R})$ convex, there exists G_{jl} such that for any $R > 0$ and for all $u, v, u', v' \in B_R(0)$ we have*

$$| G_{jl}(u,v) - G_{jl}(u',v') | \leq c(R)\, h\, (| u - u' | + | v - v' |), \tag{24}$$

$$G_{j,\alpha(j,l)}(u,v) = -G_{\alpha(j,l),j}(v,u), \tag{25}$$

$$G_{j,\alpha(j,l)}(u,u) = \nu_{j,\alpha(j,l)} F(u), \tag{26}$$

where $F(u) := \begin{pmatrix} F_1(u) \\ F_2(u) \end{pmatrix}$ *such that* $F_i'(u) = U'(u) f_i'(u), \quad i = 1,2$. *Furthermore let us assume that the conditions (4), (5), (19) are satisfied and*

$$\| u_h(.,t) \|_{L^1(\mathbf{R}^2)} \leq M \quad \text{uniformly in} \quad h. \tag{27}$$

Moreover let there exists $\beta \in [0,1[$ such that for all $K \subset\subset \mathbf{R}^2$

$$\Delta t^\beta h \sum_n \sum_{j, T_j \cap K \neq \emptyset} \sum_{l=1}^{k} | g_{jl}(u_j, u_{jl}) - \nu_{jl} f(u_j) | \leq C_2(K) \quad \text{uniformly in } \Delta t, h, \tag{28}$$

$$\Delta t^\beta h \sum_n \sum_{j, T_j \cap K \neq \emptyset} \sum_{l=1}^{k} | G_{jl}(u_j, u_{jl}) - \nu_{jl} F(u_j) | \leq C_3(K) \quad \text{uniformly in } \Delta t, h \tag{29}$$

$$U(u_j^{n+1}) - U(u_j^n) \leq -\frac{\Delta t}{| T_j |} \sum_{l=1}^{k} G_{jl}(u_j^n, u_{jl}^n). \tag{30}$$

Then

$$u_h \to u \quad \text{weak } *, \quad u \in L^\infty(\mathbf{R}^2 \times \mathbf{R}^+) \tag{31}$$

and u is the uniquely determined solution in the sense of Kruzkov of the initial value problem (1), (2).

Using the notation and assumptions introduced in §2 we will assume for the remaining part of this paper that the numerical flux $g_{jl} \in C^1(\mathbf{R}^2)$ satisfies the conditions (7), (8), (9) and the numerical entropy flux $G_{jl} \in C^1(\mathbf{R}^2)$ satisfies the conditions (24), (25), (26). Without loss of generality we also assume that

$$f_1(0) = f_2(0) = 0. \tag{32}$$

Additionally we assume the monotonicity of the scheme (6) (see (21)). Now it is easy to show that the condition (28) implies the condition (29) for a suitably chosen numerical entropy flux. This is the statement of the following lemma.

LEMMA 3.2 *Assume that for the numerical flux g_{jl} there exists a $\beta \in [0,1[$, such that for all $K \subset\subset \mathbf{R}^2$ condition (28) is satisfied. Let G_{jl} be the numerical entropy flux corresponding to the entropy (U, F_1, F_2) such that an additional compatibility condition*

$$\frac{\partial G_{jl}}{\partial v}(u,v) = U'(v)\frac{\partial g_{jl}}{\partial v}(u,v) \tag{33}$$

is fulfilled for all $u,v \in \mathbf{R}$. Moreover we suppose that

$$g_{jl} \text{ is monotonely nonincreasing in the second variable} \tag{34}$$

(which is fulfilled by the monotone schemes). Then for any $K \subset\subset \mathbf{R}^2$ we have (29).

For the Engquist-Osher- and the Lax-Friedrichs scheme we can show that the condition (33) is satisfied. Without loss of generality we assume that $F(0) = 0$. We choose the numerical entropy flux in the case of the Engquist-Osher scheme

$$G_{jl}^{EO}(u,v) := | S_{jl} | \int_0^u U'(s) \max\left(c'_{jl}(s), 0\right) ds + | S_{jl} | \int_0^v U'(s) \min\left(c'_{jl}(s), 0\right) ds\,,$$

and in the case of the Lax-Friedrichs scheme

$$G_{jl}^{LF}(u,v) := \frac{1}{2}[\nu_{jl}F(u) + \nu_{jl}F(v)] - \frac{1}{2\lambda_{jl}}\left(U(v) - U(u)\right)\,.$$

Then these numerical entropy fluxes satisfy the conditions (24), (25), (26) and the compatibility condition (33).

Next we shall consider the condition (28) of Theorem (3.1). The following lemma contains a sufficient condition for (28).

LEMMA 3.3 *Let $u_0 \in L^1(\mathbf{R}^2) \cap L^\infty(\mathbf{R}^2)$, $\| u_0 \|_{L^\infty} \leq M$ and let u_j^n be defined by the numerical scheme (6) with numerical fluxes g_{jl}, G_{jl} chosen in such a way that the conditions (33), (34) are fulfilled. Moreover, choose Δt such that the CFL-like condition*

$$\frac{\Delta t}{h} \leq \frac{1}{(k+2)\cdot c(M)\cdot c_V} \tag{35}$$

holds. Here $c(M)$ is the constant of Lipschitz continuity of g_{jl} and G_{jl} (see (7) and (24)) and c_V is defined in (5). Then there exists a $\beta \in]\frac{1}{2}, 1[$ such that for all $K \subset\subset \mathbf{R}^2$ condition (28) is satisfied.

Now it remains to consider condition (30). Therefore, we shall focus on the global control of the entropy dissipation (30) for all entropies (U, F_1, F_2) and the corresponding G_{jl}. We shall see that our compatibility condition (33) is again the crucial point in proving (30). This is contained in the next lemma and since this is the most important step we shall give the main idea of the proof.

LEMMA 3.4 *Under the assumptions of Lemma 3.3 we have (30) for all convex entropies $U \in C^2(\mathbf{R})$.*

•

PROOF: Throughout the proof we will use the notation

$$\Delta_2 G(t) := G_{jl}(u_j, t) - G_{jl}(u_j, u_j), \quad \Delta_2 g(t) := g_{jl}(u_j, t) - g_{jl}(u_j, u_j),$$

$$\partial_2 G(t) := \frac{\partial G_{jl}}{\partial t}(u_j, t), \quad \partial_2 g(t) := \frac{\partial g_{jl}}{\partial t}(u_j, t).$$

We begin by rewriting the scheme (6):

$$u_j^{n+1} = \frac{1}{k}(\sum_{l=1}^{k} \left(u_j^n - \lambda k \Delta_2 g(u_{jl})\right)) \tag{36}$$

where $\lambda = \frac{\Delta t}{|T_j|}$. The main step in the proof is the following meanvalue property of convex functions:

$$U\left(\frac{a_1 + \ldots + a_m}{m}\right) \leq \frac{1}{m}(U(a_1) + \ldots + U(a_m)) \tag{37}$$

which can be easily proved by induction. Applying this to (36) we obtain

$$U(u_j^{n+1}) \leq \frac{1}{k}\sum_{l=1}^{k} U\left(u_j^n - \lambda k \Delta_2 g(u_{jl})\right). \tag{38}$$

Hence, the proof will be finished if we prove the following inequality

$$U\left(u_j^n - \lambda k \Delta_2 g(u_{jl})\right) \leq U(u_j^n) - \lambda k \Delta_2 G(u_{jl}). \tag{39}$$

Now, setting $a := u_j^n$ and defining for all $t \in [-M, M]$

$$p(t) := U\left(a - \lambda k \Delta_2 g(t)\right) - U(a) + \lambda k \Delta_2 G(t),$$

we have $p(a) = 0$ and

$$p'(t) = U'\left(a - \lambda k \Delta_2 g(t)\right) \cdot (-\lambda k \partial_2 g(t)) + \lambda k \partial_2 G(t)$$

$$= \lambda k\, \partial_2 g(t)(U'(t) - U'(a - \lambda k \Delta_2 g(t)))$$

$$= \lambda\, k\, \partial_2 g(t)\, U''(\eta)\, (t - a + \lambda k \partial_2 g(\tilde{\eta})(t-a))$$

for some η between t and $a - \lambda k \Delta_2 g(t)$ and $\tilde{\eta}$ between a and t. Finally,

$$p'(t) = \lambda\, k\, \partial_2 g(t)\, U''(\eta)\, (t-a)\, (1 + \lambda k \partial_2 g(\tilde{\eta}))\,.$$

Since $\partial_2 g(t) \leq 0$ for all $t \in]-M, M[$ and $U'' \geq 0$ (the convexity of U) the proof can be finished by showing that $1 + \lambda k \partial_2 g(\tilde{\eta}) \geq 0$. But this is true because of the CFL-like condition (35) .

Now we can summarize the results of this paragraph in the following

THEOREM 3.5 *Let us assume that 1.3, (4), (5) and (32) are satisfied. Let $u_0 \in L^\infty(\mathbf{R}^2) \cap L^1(\mathbf{R}^2)$ and let u_j^n be defined by the numerical scheme (6) such that for g_{jl} the properties (7),(8) and (9) and for any $U \in C^2(\mathbf{R})$ convex and the coresponding G_{jl} the properties (24),(25), (26) and (33) are valid. Let the scheme (6) be monotone . Furthermore we assume the conditions (4), (5) and the CFL-like condition (35) . Then all the conditions of Theorem 3.1 are satisfied and, consequently,*

$$u_h \to u \quad weak\ *, \quad u \in L^\infty(\mathbf{R}^2 \times \mathbf{R}^+) \tag{40}$$

and u is the uniquely determined solution in the sense of Kruzkov of the initial value problem (1), (2).

PROOF: The proof is a trivial consequence of the maximum principle and Lemma 3.2, 3.3 and 3.4.

EXAMPLE 3.6 *(Engquist-Osher and Lax-Friedrichs schemes) As a consequence of the Theorem 3.5 we have in particular the convergence proof for the Engquist-Osher and the Lax-Friedrichs schemes with numerical entropy fluxes as in Example (10) and (11). Note that the convergence result obtained in [2] can be applied to the Lax-Friedrichs scheme but not to the Engquist-Osher one, since the latter doesn't satisfy the condition (2.37) of Theorem 2.3 in [2].*

REMARK 3.7 *The monotonicity condition (21) implies that the schemes for which we can prove convergence are formerly of first order. In [12] we have also shown convergence for the second order discontinuous Galerkin methods as defined in ([2]).*

4 Numerical results

In this section we shall use the finite volume (13) method on unstructured grids for the discretization of the Euler equations of gas dynamics in two and three space dimensions .

$$\partial_t u + \partial_x f_1(u) + \partial_y f_2(u) + \partial_z f_3(u) = 0\,, \tag{41}$$

where

$$u = \begin{pmatrix} \rho \\ \rho u_1 \\ \rho u_2 \\ \rho u_3 \\ e \end{pmatrix}, \qquad f_i(u) = \begin{pmatrix} \rho u_i \\ \rho u_i u_1 + \delta_{i1} p \\ \rho u_i u_2 + \delta_{i2} p \\ \rho u_i u_3 + \delta_{i3} p \\ (e+p) u_i \end{pmatrix} \tag{42}$$

and

$$p = (\gamma - 1)\left(e - \frac{\rho}{2}(u_1^2 + u_2^2 + u_3^2)\right). \tag{43}$$

We consider the supersonic flow around a sphere in $\mathbf{R}^3$ (see Figure 2). For the discretization we have used the flux-vector-splitting (13) and a grid consisting of tetrahedrons. The computations for this testproblem have been done by M. Geiben (see [7]).

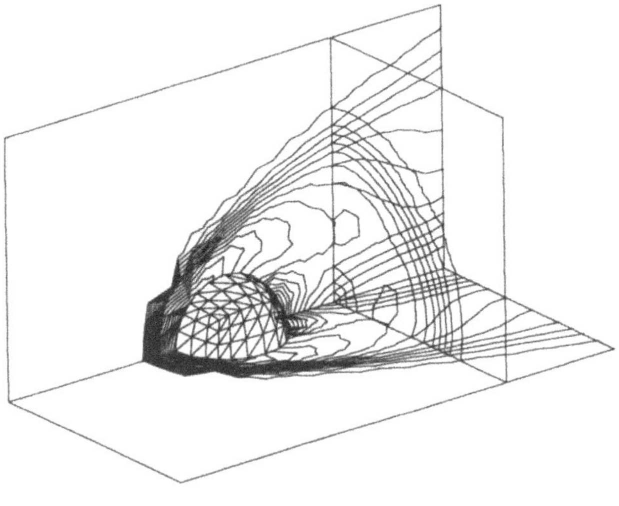

Figure 2

Finally we would like to show that it is very easy and effective to do local grid refinement in combination with grid alignment on unstructured grids.

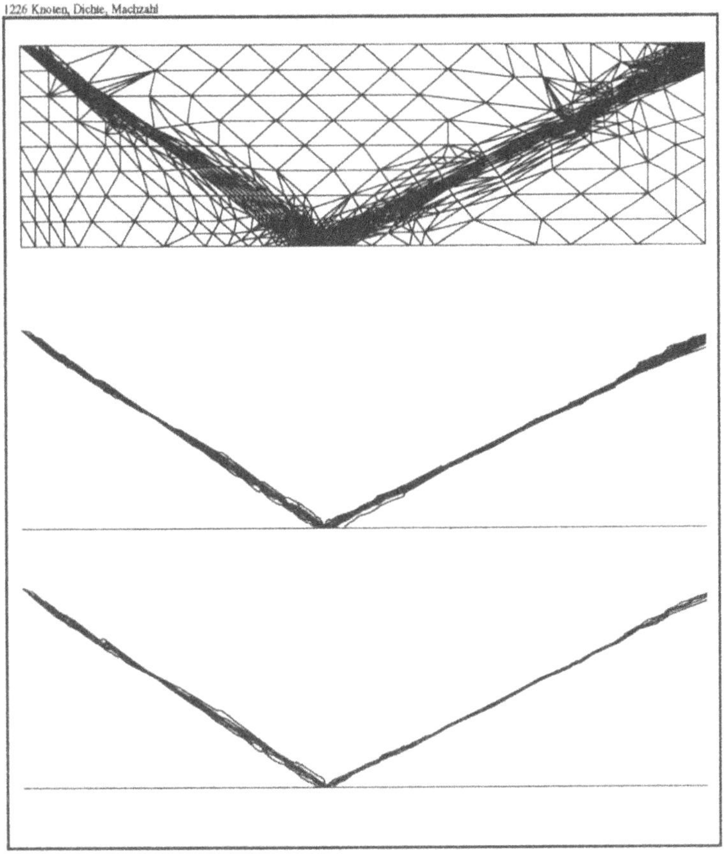

Figure 3

By this we mean an adaption of the triangles to the structure of the solution. For instance in the neighborhood of a shock the triangles could be long in the direction tangential to the shock and they should be small normal to the shock to get a better resolution. To realize this we first need a good scheme on unstructured grids as described in the sections before and second an algorithm which controls the adaption

of the triangles. Although the algorithm described before (see (6)) could be applied for this idea instead of it we have used the fluctuation-distribution scheme of Roe , Struijs and Deconinck [17] since this is formally of second order.
For the adaption of the triangle we use the algorithm of Kornhuber and Roitzsch [10] , who have used it in the context of finite element solutions for scalar, linear equations. We have combined the fluctuation-distribution scheme with the Kornhuber-Roitzsch algorithm and we have applied this to the shock reflection testproblem for the Euler equations of gas dynamics. For the local grid alignment we have used the gradients of the Machnumber.
Figure 3 shows the results of the numerical experiments for the shock reflection problem. In the three pictures we see the mesh with the local grid alignment near the shock, the contour lines of the density and the Machnumber. The grid consists of around 1300 nodes and the resolution of the shock is much better than in the case where we have only local grid refinement without grid alignment. The computations for this experiment have been done by R. Beinert.

For the visualization of the numerical data the GRAPE system has been used [8], [15], [20]. This graphic system has been developped at the Sonderforschungsbereich 256 of the Universität Bonn and at the Universität Freiburg.

References

[1] Beinert, R., Kröner, D.: Local grid alignment for the Euler equation in 2-D. Manuscript 1993.

[2] Cockburn,B. Coquel,F., LeFloch, P., Shu, C.W.: Convergence of finite volume methods. Preprint 1991.

[3] Coquel, F., LeFloch, P.: Convergence of finite difference schemes for conservation laws in several space dimensions: the corrected antidiffusive flux approach. Math. of Comp. 57, 169-210 (1991).

[4] Crandall, M.C., Majda, A.: Monotone difference approximation for scalar conservation laws. Math. of Comp. 34 (1980), 1-21.

[5] DiPerna, R.J.: Measure valued solutions of conservation laws. Arch. Rat. Mech. Anal. 88 (1985), 223-270.

[6] Engquist, B., Osher, S.: One sided difference approximations for nonlinear conservation laws. Math. of Comp. 36 (1981), 321-351.

[7] Geiben, M., Kröner, D., Rokyta, M.: A Lax-Wendroff type theorem for cell-centered, finite volume schemes in 2-D. Preprint 1993.

[8] Geiben, M., Rumpf, M.: Visualization of Finite Elements and Tools for Numerical Analysis, Report 9, SFB 256, Bonn, 1991.

[9] Johnson, C., Szepessy, A.: Convergence of a finite element method for a nonlinear hyperbolic conservation law. Math. Comput. 49 (1987), 427-444.

[10] Kornhuber, R., Roitzsch, R.: Adaptive Finite-Element Methoden für konvektionsdominierte Randwertprobleme bei partiellen Differentialgleichungen. Konrad-Zuse-Zentrum, Preprint SC 88-9, Berlin 1988.

[11] Kröner, D.: Directionally adapted upwind schemes in 2-D for the Euler equations. Priority Research Program, Results 1986-1988. Notes on numerical fluid mechanics II., Vol.25.

[12] Kröner, D., Noelle, S., Rokyta, M.: Convergence of higher order upwind finite volume schems on unstructured grids for scalar conservation laws in two space dimensions. Preprint SFB Bonn 1993.

[13] Kröner, D., Rokyta, M.:Convergence of upwind finite volume schems for scalar conservation laws in 2-D. Preprint SFB 208 (1993), Bonn ; submitted to SIAM J. of Num. Anal.

[14] Kruzkov, S.N.: First order quasilinear equations in several independent variables. Mat.Sbornik Tom 81 (123), 1970 and Math. USSR-SB 10, 217-243 (1970).

[15] Rumpf M., Schmidt A.: GRAPE, GRAphics Programming Environment, Report 8, SFB 256, Bonn, 1990.

[16] Steger, J.L., Warming, R.F.: Flux vector splitting of the inviscid gasdynamic equations with application to finite difference methods. J. Comput. Phys. 40 (1981), 263-293.

[17] Struijs, R., Deconinck, H.: Multidimensional upwind schemes for the Euler equations using fluctuation distribution. VKI, 1989/90.

[18] Szepessy, A.: Convergence of a shock capturing streamline diffusion finite element method for a scalar conservation law in two space dimensions, Math. of Comp. 53 (1989), 527-545.

[19] Vijayasundaram, G.: Transonic flow simulations using an upstream centered scheme of Godunov in finite elements. J. of Comp. Physics 63 (1986), 416-433.

[20] Wierse A., Rumpf M.: GRAPE, Eine objektorientierte Visualisierungs- und Numerikplattform. Informatik Forschung und Entwicklung 7, 145-151, 199

FROM FINITE ELEMENTS TO INTERPOLATED FINITE ELEMENTS

LIN QUN
Institute of Systems Science, Academia Sinica
Beijing, 100080, P.R. of China

Abstract

Under certain adaptive meshes, e.g. the rectangular mesh and its adaptive deformations, the finite element solution is more close to the finite element interpolation than that to the exact solution. But a higher order interpolation of the finite element solution is more close to the exact solution directly. So, the interpolated finite element solution is a better approximation to the exact solution than the finite element solution itself. The concept of interpolated finite elements makes the superconvergence to be natural since after a higher order interpolation the original lower order finite elements become a higher order element. Such an approach is a shortcut to achieve the global superconvergence up to the boundary, which is particularly useful for the eigenvalue problem. In contrast, the older superconvergence approach concentrates on finding superconvergence points and it becomes complicated when extend it up to the boundary. Therefore, interpolated finite elements renew the old superconvergence concept.

1. Basic Principle and Results

A finite element method associated with the arbitrary mesh has only a lower order of convergence. When certain special meshes are imposed on the finite element method, however, we are able to increase the convergent order either automatically or by an additional interpolation postprocess. Such kind of special meshes is the basis of our work. It includes the rectangular mesh and its adaptive generalization:

0) Rectangular mesh;

1) Deformed rectangular mesh adapted to the convex quadrilateral domain;

2) Piecewise deformed rectangular mesh adapted to the polygonal domain;

3) Mostly rectangular mesh, including the piecewise rectangular mesh mingling with boundary layers of triangular elements between adjacent pieces, adapted to the

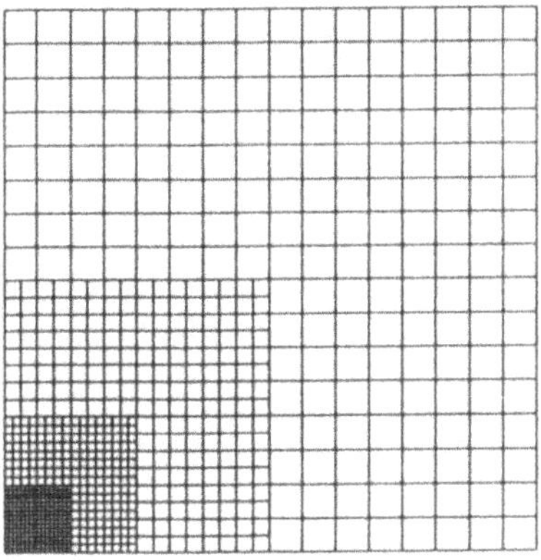

domain with the complicated boundary or reentrant corners (without knowing the singular exponent).

They are, thus, adapted to different shapes of domains. See [6] for the detailed construction.

The advantage of above meshes is making the integrals of interpolation error to be higher order, e.g. for the bilinear finite element interpolation $i_h u$ (associated with the exact solution u) and the test function φ,

0) For rectangular mesh:

$$\begin{aligned}
\int_\Omega (u - i_h u)_x \varphi_x &= O(h^2)||\varphi||_0||u||_4, \text{ if } \frac{\partial u}{\partial n} = 0 \text{ (or } \varphi = 0) \text{ on } \partial\Omega\,, \\
&= O(h^{1.5})||\varphi||_0||u||_4, \\
&= O(h^2)||\varphi||_1||u||_3\,; \\
\int_\Omega (u - i_h u)_x \varphi &= O(h^2)||\varphi||_0||u||_3, \text{ if mesh is almost uniform and } \varphi = 0 \text{ on } \partial\Omega\,, \\
&= O(h^{1.5})||\varphi||_0||u||_3, \text{ if mesh is almost uniform}, \\
&= O(h^2)||\varphi||_1||u||_3\,; \\
\int_\Omega (u - i_h u)_x \varphi_y &= O(h^2)||\varphi||_0||u||_4, \text{ if mesh is almost uniform and } \varphi = 0 \text{ on } \partial\Omega\,, \\
&= O(h)||\varphi||_0||u||_4, \text{ if mesh is almost uniform}\,, \\
&= O(h^2)||\varphi||_1||u||_3\,, \text{if} \varphi = 0 \text{ on } \partial\Omega, \\
&= O(h^2|\log h|^{1/2})||\varphi||_1||u||_4, \text{if mesh is almost uniform}\,, \\
&= O(h^{1.5})||\varphi||_1||u||_3\,.
\end{aligned}$$

1) Deformed rectangular mesh:

$$\begin{aligned}
\int_\Omega (u - i_h u)_x \varphi &= O(h^2)\|\varphi\|_0\|u\|_3, \text{ if mesh is almost uniform and } \varphi = 0 \text{ on } \partial\Omega\ , \\
&= O(h^{1.5})\|\varphi\|_0\|u\|_3, \text{ if mesh is almost uniform,} \\
&= O(h^2)\|\varphi\|_1\|u\|_3\ ; \\
\int_\Omega (u - i_h u)_x \varphi_y &= O(h)\|\varphi\|_0\|u\|_4, \text{ if mesh is almost uniform }\ , \\
&= O(h^2)\|\varphi\|_1\|u\|_3\ , \text{if} \varphi = 0 \text{ on } \partial\Omega, \\
&= O(h^2|\log h|^{1/2})\|\varphi\|_1\|u\|_4, \text{if mesh is almost uniform }\ , \\
&= O(h^{1.5})\|\varphi\|_1\|u\|_3\ .
\end{aligned}$$

2) Piecewise deformed rectangular mesh:

$$\begin{aligned}
\int_\Omega (u - i_h u)_x \varphi &= O(h^{1.5})\|\varphi\|_0\|u\|_3, \text{ if mesh is almost uniform,} \\
&= O(h^2)\|\varphi\|_1\|u\|_3; \\
\int_\Omega (u - i_h u)_x \varphi_y &= O(h)\|\varphi\|_0\|u\|_4, \text{ if mesh is almost uniform,} \\
&= O(h^2|\log h|^{1/2})\|\varphi\|_1\|u\|_4, \text{if mesh is almost uniform }\ , \\
&= O(h^{1.5})\|\varphi\|_1\|u\|_3\ .
\end{aligned}$$

3) Mostly rectangular mesh:

$$\begin{aligned}
\int_\Omega (u - i_h u)_x \varphi_x &= O(h^{1.5})\|\varphi\|_1(\|u\|_3 + \|u\|_{2,\infty}); \\
\int_\Omega (u - i_h u)_x \varphi &= O(h^{1.5})\|\varphi\|_0(\|u\|_3 + \|u\|_{2,\infty}); \\
\int_\Omega (u - i_h u)_x \varphi_y &= O(h^{1.5})\|\varphi\|_1(\|u\|_3 + \|u\|_{2,\infty}).
\end{aligned}$$

Their proofs are based on explicit integral identities introduced in [4]-[6] (see also [8]).

They are much sharper than that for an arbitrary mesh: the latter guarantees only lower order estimates, e.g.

$$\int |(i_h u - u)_x \varphi_x| = O(h)\|\varphi\|_1 = O(1)\|\varphi\|_0\ .$$

Above meshes are considered as the optimal adaptive meshes. They made, as shown in [6]-[8], the finite element solution of optimal error order. We now explain how they make the finite element gradient of superconvergence.

The integral of interpolation error characterizes the global energy error between the finite element solution u_h and the finite element interpolation $i_h u$. This is easy to see for the standard finite element scheme of Laplace equation:

$$\begin{aligned}\int_\Omega \nabla(u_h - i_h u) \nabla \varphi &= \int_\Omega \nabla(u - i_h u) \nabla \varphi \\ &= O(h^2)\|\varphi\|_1 \text{ for mesh 0) or 1)}, \\ &= O(h^{1.5})\|\varphi\|_1 \text{ for mesh 2) or 3)}, \\ &= O(h^2|\log h|^{1/2})\|\varphi\|_1 \text{ for almost uniform mesh 2)}.\end{aligned}$$

Similar sharp estimate holds true for the finite element eigensolution of Laplace operator, see Section 3.3. For the velocity of Ciarlet-Raviart or Hermann-Miyoshi scheme of biharmonic equation we refer to [9] and [11]. For the velocity of $Q_1 - Q_1$ scheme of Stokes equation we refer to [10]. We list below the global H_1-error between the bilinear finite element solution u_h (or eigensolution u^h) and the bilinear finite element interpolation.

Mesh	Δ	$\Delta u = \lambda u$	Δ^2	Stokes
	$u_h - i_h u$	$u^h - i_h u$	$u_h - i_h u$	$u_{h/2} - i_{h/2} u$
0)	h^2	h^2		
0) & a. u.			h^2	$h^{1.5}$
1)	h^2	h^2		
1) & a. u.			$h^{1.5}$	$h^{1.5}$
2)	$h^{1.5}$	$h^{1.5}$		
2) & a. u.	$h^2\|\ln h\|^{1/2}$	$h^2\|\ln h\|^{1/2}$	$h^{1.5}$	$h^{1.5}$
3)	$h^{1.5}$	$h^{1.5}$	$h^{0.75}$	$h^{1.5}$
Arbitrary	h	h		h

They are much sharper than the H_1-error between the bilinear finite element solution and the exact solution. We show below how the interpolated finite element solution provides an approximation to the exact solution with the same high order as the error listed in the above table. Let us explain such a postprocess as follows. Assume that the mesh T_h is obtained from an optimal adaptive mesh T_{2h} of size $2h$ by subdividing each element into four rectangles of size h (c.f. [1] and [16]). Let I_{2h}^2 be the biquadratic interpolation defined on T_{2h}. Since i_h is the bilinear interpolation defined on T_h it holds

$$I_{2h}^2 i_h = I_{2h}^2 \,.$$

Defining $I_{2h}^2 u_h$, the biquadratic interpolation on T_{2h} acted on the bilinear finite element solution on T_h, we have a global higher order H_1-error: e.g. for rectangular mesh,

$$I_{2h}^2 u_h - u = I_{2h}^2(u_h - i_h u) + I_{2h}^2 u - u = O(h^2)\| u \|_3 .$$

Thus, the interpolated finite element solution makes the error of one order higher than the finite element solution itself.

We list below the global H_1- errors of the interpolated finite element solution $I_{2h}^2 u_h$ (or eigensolution $I_{2h}^2 u^h$) for different meshes:

Mesh	Δ	$\Delta u = \lambda u$		Δ^2	Stokes
	$I_{2h}^2 u_h - u$	$I_{2h}^2 u^h - u$	$\frac{\| I_{2h}^2 u^h \|_1^2}{\| I_{2h}^2 u^h \|_0^2} - \lambda$	$I_{2h}^2 u_h - u$	$I_h^2 u_{h/2} - u$
0)	h^2	h^2	h^4		
0) & a. u.				h^2	$h^{1.5}$
1)	h^2	h^2	h^4		
1) & a. u.				$h^{1.5}$	$h^{1.5}$
2)	$h^{1.5}$	$h^{1.5}$	h^3		
2) & a. u.	$h^2\|\ln h\|^{1/2}$	$h^2\|\ln h\|^{1/2}$	$h^4\|\ln h\|$	$h^{1.5}$	$h^{1.5}$
3)	$h^{1.5}$	$h^{1.5}$	h^3	$h^{0.75}$	$h^{1.5}$
Arbitrary	h	h	h^2		h

They are much sharper than the H_1−errors of the finite element solution itself.

Interpolated finite elements change the original finite element shape function to the interpolated shape function, i.e. the quadratic interpolation of the bilinear shape function. The latter is considered as an optimal shape function. Therefore, the interpolated shape function based on optimal adaptive meshes causes the superconvergence.

2. Extension

1. High order elements

The principle and results in Section 1 can be extended to finite elements of degree of $p \geq 2$, particularly to biquadratic finite elements, on the rectangular shape domain. Roughly speaking, the interpolated finite element solution makes the error of two order higher than the finite element solution itself. This has been done in [13].

The sharp estimates in Section 1 are originated from an explicit identity for the integral of interpolation error. In the bilinear case, such an interpolation is no difference with the usual nodal interpolation. The analogy for high order case, however, is non-trivial. The problem is that, in the high order case, the traditional nodal interpolation does not give an explicit integral identity (and higher superconvergence). In order to recover the explicit integral identity we have to forsake the nodal interpolation and choose the "vertices-edges-element" interpolation $i_h^p u$ defined as follows.

Let the rectangular mesh

$$T_h = \{\tau\},$$

where (x_τ, y_τ) denotes the center of the element τ and $2h_\tau, 2k_\tau$ are its widths in the x- and y- direction, respectively. Further, we set

$$d_\tau = \max\{h_\tau, k_\tau\},$$

$$h = \max\{d_\tau : \tau \in T_h\},$$

and denote the vertices and edges of τ by p_i and l_j, respectively. Let

$$S_h^p = \{v \in H_0^1(\Omega) : v \in Q^p(\tau), \tau \in T_h\},$$

$$Q^p(\tau) = \text{span}\{x^i y^j : 0 \le i, j \le p\}.$$

The interpolation operator $i_h^p \in (C \to S_h^p)$ is defined by the following "vertices-edges-element" conditions:

$$i_h^p u(p_i) = u(p_i),$$

$$\int_{l_j} (i_h^p u - u)v = 0, \quad \forall v \in P^{p-2}(l_j),$$

$$\int_\tau (i_h^p u - u)v = 0, \quad \forall v \in Q^{p-2}(\tau),$$

where $P^{p-2}(l_j)$ and $Q^{p-2}(\tau)$ are the polynomial space of degree $p-2$ on l_j and the polynomial space of degree $p-2$ on τ, respectively. Since

$$i_h^p u \in S_h^p \subset C, \quad \forall u \in C:$$

$$i_h^p v = v, \quad \forall v \in S_h^p$$

we have, by the Bramble-Hilbert lemma,

$$\|i_h^p u - u\|_{r,q} \le ch^{p+1-r}\|u\|_{p+1,q}, \quad 1 \le q \le \infty.$$

(if T_h is almost regular when $r > 0$. See Ciarlet [2]).

Such an interpolation is different with the nodal interpolation but can also be constructed with the similar basis $\{\varphi_i\}$ such that

$$i_h^p u = \sum \hat{u}_{ij}\varphi_i(\zeta)\varphi_j(\eta), \quad \text{on } \tau.$$

For example $p = 2$, we have

$$\varphi_1(\alpha) = \frac{1}{4}(3\alpha + 1)(\alpha - 1),$$

$$\varphi_2(\alpha) = \frac{3}{4}(1 - \alpha^2),$$

$$\varphi_3(\alpha) = \frac{1}{4}(3\alpha - 1)(\alpha + 1),$$

$$\begin{aligned}
\xi &= \frac{x - x_\tau}{h_\tau}, \eta = \frac{y - y_\tau}{k_\tau}, \\
\hat{u}_{11} &= u(x_\tau - h_\tau, y_\tau - k_\tau), \\
\hat{u}_{21} &= \int_{x_\tau - h_\tau}^{x_\tau + h_\tau} u(x, y_\tau - k_\tau)\, dx/h_\tau, \\
u_{31} &= u(x_\tau + h_\tau, y_\tau - k_\tau), \\
\hat{u}_{12} &= \int_{y_\tau - k_\tau}^{y_\tau + k_\tau} u(x_\tau - h_\tau, y)\, dy/k_\tau, \\
\hat{u}_{22} &= \int_\tau u(x, y) dx dy / h_\tau k_\tau, \\
\hat{u}_{32} &= \int_{y_\tau - k_\tau}^{y_\tau + k_\tau} u(x_\tau + h_\tau, y) dy/k_\tau, \\
\hat{u}_{13} &= u(x_\tau - h_\tau, y_\tau + k_\tau), \\
\hat{u}_{23} &= \int_{x_\tau - h_\tau}^{x_\tau + h_\tau} u(x, y_\tau + k_\tau) dx/h_\tau, \\
\hat{u}_{33} &= u(x_\tau + h_\tau, y_\tau + k_\tau).
\end{aligned}$$

We now can see how the "vertices-edges-element" interpolation makes the explicit integral identity to be possible. Such an identity will use the interpolation error functions

$$F(y) = \frac{1}{2}((y - y_\tau)^2 - k_\tau^2),$$

$$E(x) = \frac{1}{2}((x - x_\tau)^2 - h_\tau^2).$$

Note that

$$(F^2(y))^{(3)} = 6(y - y_\tau),$$

$$(F^3(y))^{(4)} = 45(y - y_\tau)^2 - 9k_\tau^2.$$

Together with the integration by parts, we have

Lemma 1. If

$$\int_{l_i} w\,dx = 0, \quad i = 1,2,$$

then

$$\int_\tau (y - y_\tau)w = -\frac{1}{6}\int_\tau F^2(y)\,w_{yyy},$$

where l_i denote the edges of τ parallel to x-axis. If

$$\int_{l_i} w\,dx = \int_\tau w = 0,$$

then

$$\int_\tau (y - y_\tau)^2 w = \frac{1}{45}\int_\tau F^3(y)\,w_{yyyy}.$$

Having the Lemma 1, we have the following explicit integral identity.
Lemma 2. Under the general rectangular meshes, for $v \in S_h^2$,

$$\begin{aligned}(\nabla(i_h^2 u - u), \nabla v) &= \frac{1}{6}\sum\int_\tau (F^2(y)u_{xxyyy}(v_y + (y_\tau - y)v_{yy}) \\ &\quad + E^2(x)u_{xxxyy}(v_x + (x_\tau - x)v_{xx}) \\ &\quad -\frac{1}{15}F^3(y)u_{xyyyy}v_{xyy} - \frac{1}{15}E^3(x)u_{xxxxy}v_{xxy}),\end{aligned}$$

i.e. the integral of interpolation error is expressed by the error functions F and E.

Proof. Set

$$W = i_h^2 u - u.$$

Since

$$v(x,y) = v(x,y_\tau) + (y - y_\tau)v_y(x,y_\tau) + \frac{1}{2}(y - y_\tau)^2 v_{yy}(x,y),$$

we have

$$\int_\tau W_x v_x = \int_\tau W_x v_x(x,y_\tau) + \int_\tau (y - y_\tau)W_x v_{xy}(x,y_\tau) + \frac{1}{2}\int_\tau (y - y_\tau)^2 W_x v_{xyy}.$$

The "vertices-edges-element" conditions for i_h^2 are now used to reduce the following integrals:

$$\begin{aligned}\int_\tau W_x v_x(x,y_\tau) &= -\int_\tau W v_{xx}(x,y_\tau) \\ &\quad +\left(\int_{s_1} - \int_{s_2}\right) W v_x(x,y_\tau)\,dy = 0, \\ \int_{l_i} W_x v_{xy}(x,y_\tau)\,dx &= -\int_{l_i} W v_{xxy}(x,y_\tau)dx = 0, \\ \int_{l_i} W_x v_{xyy}\,dx &= \int_\tau W_x v_{xyy} = 0,\end{aligned}$$

where s_i denote the edges of τ parallel to y-axis. Hence, substituting w in Lemma 1 by $W_x v_{xy}(x, y_\tau)/$ or $W_x v_{xyy}$, we obtain

$$\begin{aligned}\int_\tau (y - y_\tau) W_x v_{xy}(x, y_\tau) &= \frac{1}{6}\int_\tau F^2(y) u_{xyyy} v_{xy}(x, y_\tau) \\ &= -\frac{1}{6}\int_\tau F^2(y) u_{xxyyy} v_y(x, y_\tau) \\ &\quad +\frac{1}{6}\left(\int_{s_1} - \int_{s_2}\right) F^2(y) u_{xyyy} v_y(x, y_\tau)\, dy, \qquad (6)\end{aligned}$$

$$\int_\tau (y - y_\tau)^2 W_x v_{xyy} = -\frac{1}{45}\int_\tau F^3(y) u_{xyyyy} v_{xyy}.$$

Since $F^2(y) u_{xyyy} v_y(x, y_\tau)$ in (6) is continuous on s_i and $v_y = 0$ on the boundaries parallel to y-axis, the line integrals in (6) are cancelled during the summation of $\tau \in T_h$. And, $v_y(x, y_\tau)$ can be replaced by

$$v_y(x, y_\tau) = v_y(x, y) + (y_\tau - y) v_{yy}(x, y).$$

Similar identity holds for $\int_\tau W_y v_y$ and hence Lemma 2 is proved.

Let $R_h u \in S_h^p$ be the Ritz projection of Laplace operator.

With the inverse inequality, Lemma 2 implies a sharper estimate:

$$(\nabla(i_h^2 u - R_h u), \nabla v) = O(h^4)\|u\|_{5,q}\|v\|_{1,q'}, \quad \forall v \in S_h^2.$$

In particular, we have, by a standard argument ([17]),

Proposition 1. Under the general rectangular meshes,

$$\|i_h^2 u - R_h u\|_{1,q} \le c h^4 \|u\|_{5,q}.$$

Thus, the "vertices-edges-element" interpolation is more close to the finite element solution than the standard nodal interpolation: the latter gives only an order of $O(h^3)$. This is the reason that we prefer to the nonstandard interpolation rather than the standard interpolation.

Lemma 1, 2 and Proposition 1 can be generalized to $p \ge 3$.

Lemma 3. If

$$\begin{aligned}\int_{l_i} w\, dx &= 0, \quad i = 1, 2, \\ \int_\tau y^j w &= 0, \quad 0 \le j \le m - 2,\end{aligned}$$

then

$$\int_\tau (y - y_\tau)^m w = (-1)^m 2^{m+1} \frac{m!}{(2(m+1))!}\int_\tau F^{m+1}(y)\partial_y^{m+2} w.$$

Lemma 4. Under the general rectangular meshes, for $v \in S_h^p$,

$$\int_\Omega \partial_x(i_h^p u - u)\partial_x v = -(-2)^p \frac{1}{(2p)!}\sum \int_\tau F^p(y)\partial_x^2\partial_y^{p+1}u(\partial_y^{p-1}v + (y_\tau - y)\partial_y^p v)$$
$$+(-2)^{p+1}\frac{1}{(2(p+1))!}\sum \int_\tau F^{p+1}(y)\partial_x\partial_y^{p+2}u\partial_x\partial_y^p v,$$

and a similar identity for $\int_\tau \partial_y(i_h^p u - u)\partial_y v$.

Proof. Let

$$W = i_h^p u - u.$$

Then

$$\int_\tau W_x v_x = \sum_{j=0}^{p}\frac{1}{j!}\int_\tau (y - y_\tau)^j W_x \partial_x \partial_y^j v(x, y_\tau),$$

where the first $p-2$ terms disappear:

$$\sum_{j=0}^{p-2}\frac{1}{j!}\int_\tau (y - y_\tau)^j W_x \partial_x \partial_y^j v(x, y_\tau)$$
$$= -\sum_{j=0}^{p-2}\frac{1}{j!}\int_\tau (y - y_\tau)^j W \partial_x^2 \partial_y^j v(x, y_\tau) = 0,$$

and the last two terms can be calculated by Lemma 3 :

$$\frac{1}{p!}\int_\tau (y - y_\tau)^p W_x \partial_x \partial_y^p v = (-2)^{p+1}\frac{1}{(2(p+1))!}\int_\tau F^{p+1}(y)\partial_x\partial_y^{p+2}u\partial_x\partial_y^p v,$$

$$\frac{1}{(p-1)!}\int_\tau (y - y_\tau)^{p-1} W_x \partial_x \partial_y^{p-1} v(x, y_\tau)$$
$$= (-1)^p 2^p \frac{1}{(2p)!}\int_\tau F^p(y)\partial_x\partial_y^{p+1}u\partial_x\partial_y^{p-1}v(x, y_\tau)$$
$$= -(-2)^p \frac{1}{(2p)!}\int_\tau F^p(y)\partial_x^2\partial_y^{p+1}u\partial_y^{p-1}v(x, y_\tau)$$
$$+(-2)^p \frac{1}{(2p)!}\left(\int_{s_1} - \int_{s_2}\right)F^p(y)\partial_x\partial_y^{p+1}u\partial_y^{p-1}v(x, y_\tau)dy,$$

the line integrals on s_i are cancelled during the summation of $\tau \in T_h$. Lemma 4 is proved.

Lemma 4 implies a sharp estimate: for $v \in S_h^p$,

$$(\nabla(i_h^p u - R_h u), \nabla v) = O(h^{p+3-r})\|u\|_{p+3,q}\|v\|_{2-r,q'}, r = 0, 1.$$

In particular, we have

Proposition 2. Under the general rectangular meshes,

$$\|i_h^p u - R_h u\|_{r,q} \le ch^{p+3-r}\|u\|_{p+3,q}, \quad r = 0, 1.$$

Remark. Although the " vertices - edges - element" interpolation $i_h^p u$ is already very closed to $R_h u$ we wonder if there exists another interpolation to approximate to $R_h u$ with, say, $2p$ order.

In order to approximate u with the same high order as that in propositions 1 and 2, we will use again the interpolation postprocess I_{2h}^{p+2}.

Let us observe first the 1-d case, where I_{2h}^{p+2} is determined by the following " vertices-intervals" conditions:

$$I_{2h}^{p+2}u(p_i) = u(p_i), \quad 1 \le i \le 3,$$

$$\int_{l_j} I_{2h}^{p+2}u = \int_{l_j} u, \quad 1 \le j \le 2,$$

$$\int_L I_{2h}^{p+2}u\, v = \int_L u\, v, \quad \forall v \in P^{p-2}(L),$$

with $L = l_1 \cup l_2 \in T_{2h}, l_j \in T_h$ and the vertices p_i of l_j. Moreover, $I_{2h}^{p+2}u$ can also be constructed by the basic functions $\{\varphi_i\}$:

$$\begin{cases} \varphi_i(p_j) = \delta_{i,j}, & 1 \le j \le 3, \quad 1 \le i \le 3, \\ \int_{l_j} \varphi_i = 0, & 1 \le j \le 2, \quad 1 \le i \le 3, \\ \int_L \varphi_i v = 0, & \forall v \in P^{p-2}(L), \quad 1 \le i \le 3, \end{cases}$$

$$\begin{cases} \varphi_i(p_j) = 0, & 1 \le j \le 3, \quad 4 \le i \le 5, \\ \int_{l_j} \varphi_i = \delta_{i-3,j}, & 1 \le j \le 2, \quad 4 \le i \le 5, \\ \int_L \varphi_i v = 0, & \forall v \in P^{p-2}(L) \backslash R^1, \quad 4 \le i \le 5, \end{cases}$$

$$\begin{cases} \varphi_i(p_j) = 0, & 1 \le j \le 3, \quad 6 \le i \le p+3, \\ \int_{l_j} \varphi_i = 0, & 1 \le j \le 2, \quad 6 \le i \le p+3, \\ \int_\tau (x-p_2)^j \varphi_i = \delta_{i-5,j}, & 1 \le j \le p-2, \quad 6 \le i \le p+3, \end{cases}$$

where $p_2 = l_1 \cap l_2$. For example $p = 2$: for $L \in T_{2h}$,

$$\begin{aligned}
\varphi_1(\alpha) &= \frac{1}{4}\alpha(\alpha-1)(5\alpha^2+\alpha-2),\\
\varphi_2(\alpha) &= \frac{1}{4}\alpha(1-\alpha^2)(15\alpha-8),\\
\varphi_3(\alpha) &= (\alpha^2-1)(5\alpha^2-1),\\
\varphi_4(\alpha) &= \frac{1}{4}\alpha(1-\alpha^2)(15\alpha+8),\\
\varphi_5(\alpha) &= \frac{1}{4}\alpha(\alpha+1)(5\alpha^2-\alpha-2),
\end{aligned}$$

where

$$\alpha = \frac{x-p_2}{h}, \quad h = \begin{cases} |p_3-p_2|, & x \in (p_2,p_3),\\ |p_2-p_1|, & x \in (p_1,p_2). \end{cases}$$

$I_{2h}^{p+2}u$ in 2-d case is constructed by tensor products $\{\varphi_i(x)\varphi_j(y): \quad 1 \le i,j \le p+3\}$:

$$I_{2h}^{p+2}u = \sum \hat{u}_{ij}\varphi_i(x)\varphi_j(y),$$

where $\hat{u}_{ij}$ are the values of u on the vertices p_i of the four subrectangles $\tau_i \in T_h$ and some integral mean values of u with weight $(x-x_0)^i(y-y_0)^j$ on some edges l_j of τ_i and elements τ_i, where $(x_0,y_0) = \cap_1^4\tau_i$.

The following properties of I_{2h}^{p+2} can be checked:

$$I_{2h}^{p+2} i_h^p = I_{2h}^{p+2},$$

$$\|I_{2h}^{p+2} v\|_{r,q} \le c\|v\|_{r,q}, v \in S_h^p,$$

$$\|I_{2h}^{p+2} u - u\|_{r,q} \le ch^{p+3-r}\|u\|_{p+3,q}.$$

We have then

$$I_{2h}^{p+2} R_h u - u = I_{2h}^{p+2}(R_h u - i_h^p u) + I_{2h}^{p+2} u - u$$

and hence the error of interpolated finite element solution has the same high order as that in propositions 1 and 2, i.e. two order higher than the finite element solution itself. In particular, when $p = 2$, there holds

$$\|I_{2h}^4 R_h u - u\|_{1,\infty} \le ch^4.$$

2. The 3-d case

This has been done in [12].

3. Eigenvalue problem

The global superconvergence of interpolated finite elements is particularly useful for the eigenvalue problem.

Let (λ, u) and (λ_h, u_h) denote the exact eigenpair of Laplace operator and the associated approximate eigenpair in the finite element space S_h^p, respectively. Again, $R_h u \in S_h^p$ be the Ritz projection of Laplace operator.

Lemma 5. There holds, for $p \geq 2$,

$$\|u_h - R_h u\|_{1,2} \leq ch^{p+2},$$

i.e. the finite element eigensolution is more close to the Ritz projection than that to the exact eigensolution.

Proof. Let $K = (-\Delta)^{-1}$. By the identity

$$\begin{aligned}(I - \lambda K)(R_h u - u_h) &= (\lambda_h R_h K - \lambda K)(R_h u - u_h) \\ &\quad +(\lambda - \lambda_h) R_h K u - \lambda_h R_h K(R_h u - u).\end{aligned}$$

We have

$$(I - \lambda K)(R_h u - u_h - (\nabla(R_h u - u_h), \nabla u)u) = r_h,$$

where

$$r_h = (\lambda_h R_h K - \lambda K)(R_h u - u_h) + (\lambda - \lambda_h) R_h K u - \lambda_h R_h K(R_h u - u).$$

Note that

$$|\lambda - \lambda_h| \leq ch^{2p}, \quad \|R_h u - u\|_{-1,2} \leq ch^{p+2}.$$

Hence,

$$\|r_h\|_{1,2} \leq ch^{p+2} + ch\|R_h u - u_h\|_{1,2},$$

and

$$\begin{aligned}\|R_h u - u_h\|_{1,2} &\leq \|r_h\|_{1,2} + c|(\nabla(R_h u - u_h), \nabla u)| \\ &\leq ch\|R_h u - u_h\|_{1,2} + ch^{p+2} \\ &\quad +|(\nabla(R_h u - u_h), \nabla u)|.\end{aligned}$$

On the other hand, there holds

$$\begin{aligned}(\nabla(R_h u - u_h), \nabla u) &= \lambda(R_h u - u, u) + \lambda(u - u_h, u) \\ &= (\nabla(R_h u - u), \nabla(R_h u - u)) + \frac{\lambda}{2}\|u - u_h\|_{0,2}^2,\end{aligned}$$

and Lemma 5 is proved.

Lemma 6. For $w \in H_0^1(\Omega)$ and $w \neq 0$, there hods

$$\left| \frac{\| \nabla w\|_{0,2}^2}{\|w\|_{0,2}^2} - \lambda \right| \leq c \frac{\| \nabla (w-u)\|_{0,2}^2}{\|w\|_{0,2}^2}.$$

See [15].

Note that

$$I_{2h}^{p+2} u_h - u = I_{2h}^{p+2}(u_h - R_h u) + i_{2h}^{P+2} R_h u - u.$$

And hence the error of interpolated finite element eigensolution has the same high order as that in Lemma 5. Setting

$$w = I_{2h}^{p+2} u_h$$

we obtain an order of $O(h^{2p+4})$ for the eigenvalue approximation, i.e. the interpolated finite element eigensolution makes the eigenvalue approximation of four order higher than the finite element eigensolution itself: the latter gives only an order of $O(h^{2p})$. In particular, when $p = 2$, there holds

$$\frac{\| I_{2h}^2 u^h \|_1^2}{\| I_{2h}^2 u^h \|_0^2} - \lambda = O(h^4).$$

So far, we discussed only the Laplace operator. The same sharp estimates hold also true for the biharmonic equation, see [11] for details.

4. Extrapolation and Defect correction

They have been done in [12] and [13].

In sum, they are two concepts used in the paper, the optimal adaptive meshes and the interpolated finite elements. We hope that these concepts can play as the tools to recover the error order for the finite element solution and the finite element gradient for various problems.

References

1. H. Blum, Defect correction schemes for mixed and nonconforming finite elements, *Bonn. Math. Schriften*, Nr. 228 (1991).
2. Ciarlet, Ph.G.: The Finite Element Method for Elliptic Problems, North-Holland, 1976.
3. Krizek, M., Neittaanmaki, P.: On Superconvergence techniques, *Acta Appl. Math.* 9(1987), 175–198.
4. Lin, Q.: An integral identity and interpolated postprocess in superconvergence, Research Report 90–07, Institute of Systems Science, Academia Sinica (1990), 1-6.
5. Q. Lin, A rectangle test for finite element analysis, Proceedings of Systems Science & Systems Engineering, Great Wall (H.K.) Culture Publish Co. (1991), 213-216.

6. Q. Lin, From rectangle test to optimal adaptive meshes, ibid, 247-251.
7. Q. Lin, Global error expansion and superconvergence for higher order interpolation of finite elements, J. Comput. Math., Supplementary Issue (1992), 286-289.
8. Q. Lin, What is the optimal mesh, the First China-Japan Joint Seminar on Numerical Mathematics, August 1992, Beijing, to appear in WORLD SCIENTIFIC PUBLISHING CO. PTE. LTD.(1993).
9. Q. Lin, J. Li and A. Zhou, A rectangle test for Ciarlet-Raviart and Herrmann-Miyoshi schemes, ibid, 230-233.
10. Q. Lin, J. Li and A. Zhou , A rectangle test for Stokes equations, ibid, 240-241.
11. Q. Lin and M. Liu, A rectangle test for biharmonic problem, ibid, 238-239.
12. Q. Lin and N. Yan, A rectangle test in R^3, ibid, 242-246.
13. Q. Lin , N. Yan and A. Zhou, A rectangle test for interpolated finite elements, ibid, 217-229.
14. Q. Lin and A. Zhou, A rectangle test for first order hyperbolic equation, ibid, 234-235.
15. Q. Lin and Y. Yang, Interpolation and correction of finite elements, *Math. in Practice and Theory*, 3(1991), 29-35.
16. R. Rannacher, Defect correction for the finite element method, Metz Days on Numer. Anal., Univ. Mets. (June 1990), to appear.
17. Zhu, Q.D., Lin, Q.: Superconvergence theory in the finite element method, Hunan Science Press, 1989.

Blow-up for Perturbed Conservation Laws

R. NATALINI

Istituto per le Applicazioni del Calcolo "M. Picone"
Viale del Policlinico 137, 00161 Roma, Italy

A. TESEI

Dipartimento di Matematica "G. Castelnuovo"
Università di Roma "La Sapienza", P.zza A. Moro
00185 Roma, Italy

ABSTRACT

We investigate the behaviour of nonnegative entropy solutions to the Cauchy problem for a first order quasilinear equation with a superlinear source term. The occurence of blow-up phenomena and the continuation after the blow-up time are studied.

1. Introduction

We study the Cauchy problem

$$\begin{cases} \partial_t u + \partial_x(\frac{1}{m}u^m) = u^p & \text{in } (0,\infty)\times \mathbf{R} \\ u = u_0 & \text{in } \{0\}\times \mathbf{R} \ , \end{cases} \tag{1.1}$$

where u_0 is a given bounded nonnegative function with compact support and $m > 1$, $p > 1$.

Under the previous assumptions there exists a unique local entropy solution of (1.1), which is nonnegative and bounded (see Section 2). Denote by $(0, T_M)$, with $T_M = T_M(u_0) > 0$, its maximal interval of existence. If $T_M < +\infty$, then T_M is called the *blow-up time* (in the global supremum norm) of the solution u.

In this note we shall describe the behaviour of the solutions of (1.1), depending on different values of m, p and the size of the initial data. In fact, if $p \leq m+1$, then $T_M < +\infty$ for any positive initial data; if $p > m+1$, solutions could be global for small initial data or blow up for large ones. Observe that similar phenomena arise for a wide class of nonlinear (degenerate) parabolic problems with superlinear source term; see [SGKM], [Ug] and references therein.

The mechanism of blow-up is very different when $p \leq m$ or $p > m$. In the former case, solutions of (1.1) remain locally bounded as $t \to T_M^-$. Then blow-up of the global supremum norm occurs since solutions grow up while their support expands at infinity as $t \to T_M^-$. On the contrary if $p > m$, and for $p > m+1$ if initial data are large enough, solutions become locally unbounded as $t \to T_M^-$.

When blow-up occurs, it is interesting to discuss whether the solutions can be continued beyond the blow-up time in some suitable regions. This is possible on the whole spatial real line if $p \leq m$ (see [Na] and Section 3). The same is true in some half-line $(-\infty, x_{BU}(t))$ if $p > m$, $t \geq T_M$. For simple initial data the set $\{(t,x) | \, t \geq T_M, \, x = x_{BU}(t)\}$ is a regular convex curve, called the *blow-up curve* of u. In any case, solutions can be continued beyond the blow-up time for a large class of initial data (for example, bounded measurable functions with compact support). This is in sharp contrast with the situation encountered in parabolic problems [BC].

The note is organized as follows: in Section 2 we state the local existence and uniqueness theorem for bounded entropy solutions of (1.1). As is well known, since solutions will be in general discontinuous, some admissibility conditions (called the entropy conditions) are needed to select the unique meaningful solution [La], [Ol], [Kr]. Section 3 is devoted to results concerning the *interaction problem* – namely, problem (1.1) when u_0 is the characteristic function of a bounded interval.

Finally, results for general initial data can be obtained by comparison methods from those of section 3. Details and proofs can be found in [NT1] (also [Na] for $p \leq m$). A general approach to multidimensional (scalar) conservation laws with source term will be considered in [NT2].

2. Preliminaries

Denote by Σ_T the strip $\{(t,x) | \, t \in (0,T], x \in \mathbf{R}\}$ $(T > 0)$; set $\Sigma := \Sigma_\infty$. In the sequel we will consider for (1.1) only locally bounded entropy solutions (see [Kr]). It is easy to extend the definition of entropy solution to any connected region Q such that $\Sigma_T \subseteq Q \subseteq \Sigma$, for some $T > 0$.

Concerning local existence and uniqueness of entropy solutions to (1.1), the following result holds (see [NT2]).

THEOREM 2.1. *For any nonnegative function* $u_0 \in L^\infty(\mathbf{R})$ *there exist* $T > 0$, *only depending on* $\|u_0\|_{L^\infty}$ *and an entropy solution* u *of (1.1) such that* $u \in L^\infty(\Sigma_T) \cap C([0,T]; L^1_{loc}(\mathbf{R}))$. *Moreover u is the unique bounded entropy solution of (1.1) in* Σ_T. *Either* $u \in L^\infty(\Sigma_T)$ *for any* $T > 0$, *or there exists a maximal time of life* $T_M < +\infty$ *such that*

$$\lim_{T \to T_M^-} \|u\|_{L^\infty(\Sigma_T)} = +\infty. \tag{2.1}$$

According to the above result an entropy solution of (1.1) is said to be global if $u \in L^\infty(\Sigma_T)$ for any $T > 0$. Otherwise, if $T_M < +\infty$, then T_M is called the blow-up time of the solution. In many cases it is possible to continue entropy solutions beyond the blow-up time in some suitable regions, according to the following definition (see [Na], [NT 1,2] for more general considerations).

DEFINITION 2.1. *Let* u *be the (unique) entropy solution of (1.1) in* Σ_{T_M}, *with* $T_M < +\infty$. *A measurable function* V *is called a continuation of* u *if there exists a connected region* $Q \subseteq \Sigma$ *such that:*

(i) $\Sigma_{T_M} \subseteq Q$;
(ii) $V \in L^\infty_{loc}(Q)$;
(iii) V *is an entropy solution of (1.1) in* Q;
(iv) $V = u$ *a.e. in* Σ_{T_M}.

3. Interaction waves

In this section we consider Cauchy data in (1.2) of the type

$$u_0 = \sigma \chi_{[0,l]} \qquad (\sigma, l > 0). \tag{3.1}$$

The problem (1.1), with initial data given by (3.1), will be called the *interaction problem*. The solution is constructed by the method of characteristics as a combination of elementary waves.

a) Similarity waves. An important rôle is played by the following family of elementary solutions of Eq.(1.1):

$$u(t,x) = t^{-\frac{1}{p-1}} f(\eta) \ , \quad \eta := x t^{\frac{m-p}{p-1}} \ . \tag{3.2}$$

Here f is the unique nonnegative solution of the problem:

$$\begin{cases} \left(f^{m-1} + \frac{m-p}{p-1}\eta\right) f' - f\left(\frac{1}{p-1} + f^{p-1}\right) = 0 \quad \text{in } (0,\overline{\eta}) \\ \frac{f^{m-1}}{\eta} \to 1 \quad \text{as } \eta \to 0 \ , \end{cases} \tag{3.3}$$

where

$$\overline{\eta} := \begin{cases} +\infty & \text{if } \ p \leq m \\ \frac{(p-1)^{\frac{p-m}{p-1}}}{p-m} & \text{if } \ p > m. \end{cases}$$

If $p > m$, the function f blows up as $\eta \to \overline{\eta}$. In this case an important rôle will be played by the blow-up curve $\eta(t) = \overline{\eta}$, namely

$$x_{BU}(t) := \overline{\eta} t^{\frac{p-m}{p-1}}. \tag{3.4}$$

b) Support properties. Let us denote by $x = X(t,y)$ the characteristic of (1.1) issued at $x = y$. Define also

$$U = U(t,y) := u(t, X(t,y)) \ .$$

It is easily seen that for $p \neq m$ the couple (X, U) is given by

$$U(t,y) = U_0[1 - (p-1)U_0^{p-1}t]^{-\frac{1}{p-1}} \ , \tag{3.5}$$

$$X(t,y) = y - \frac{U_0^{m-p}}{m-p}\{1 - [1 + (p-1)U_0^{p-1}t]^{-\frac{m-p}{p-1}}\} \tag{3.6}.$$

The case $p = m$ is similar. We shall denote by $x = X_0(t) := X(t,0)$ the characteristic issued at $x = 0$ and by $x = S_0(t)$ the shock issued at $x = l$. From the Rankine-Hugoniot condition, as long as S_0 does not intersect the characteristic X_0, it is possible to get the explicit expression of $S_0(t)$.

PROPOSITION 3.1. *(i) Let $p \leq m$. Then there is a unique intersection between S_0 and X_0;*
(ii) Let $p > m$. Then the shock S_0 intersects the characteristic X_0 if and only if

$$\lambda := l\sigma^{p-m} < \frac{m-1}{m(p-m)} =: M_0 \ . \tag{3.7}$$

In this case the intersection is unique.

If S_0 and X_0 intesect each other, this happens at the point

$$(x,t) = (\bar{x},\bar{t}) := (\frac{m}{m-1}l,\ X_0^{-1}(\bar{x})) \ .$$

In such case we denote by $x = S(t)$ the shock issued at $(\bar{x},\bar{t})$. It satisfies the Rankine-Hugoniot condition at the jump, where $u(S(t)+0,t) \equiv 0$ and $u(S(t)-0,t)$ is given by the similarity solution (3.2).

Concerning the support properties of the solution we have the following results.

PROPOSITION 3.2. *Let $p \leq m$. Then there exists $T > 0$ (depending on σ, l, m, p) such that*

$$\lim_{t \to T^-} S(t) = +\infty \ . \tag{3.8}$$

If $p < m$, there exist $A, B > 0$ such that

$$A \leq (T-t)^{\frac{m-p}{p-1}} S(t) \leq B \tag{3.9}$$

for any $t \in (T-\delta, T)$ ($\delta > 0$ small enough).

PROPOSITION 3.3. *(i) Let $p \in (m, m+1]$ and (3.7) hold. Then there exists $T > 0$ (depending on σ, l, m, p) such that*

$$\lim_{t \to T^-} S(t) = \bar{\eta} \ . \tag{3.10}$$

(ii) Let $p > m+1$. Then there exists $M_1 \in (0, M_0)$ (M_1 depending only on σ, l, m, p) such that

a) *if $\lambda \in (M_1, M_0)$, then (3.10) holds true for some $T > 0$;*
b) *if $\lambda \in (0, M_1]$ the shock S exists in $[\bar{t}, \infty)$.*

Remark. It follows from (3.10) that the shock S intersects the blow-up curve x_{BU} at $t = T$. Then, if $\lambda < M_1$ no blow-up occurs.

c) Solution of the Interaction problem. Now we construct the entropy solution of (1.1) with the initial condition (3.1).

Let us consider first the case $p \leq m$. Let $T > 0$ be given by Proposition 3.2; define $u : \Sigma_T \to [0, +\infty)$ as follows

$$u(t,x) := \begin{cases} t^{-\frac{1}{p-1}} f(xt^{\frac{m-p}{p-1}}) & ((t,x) \in I_1) \\ \sigma[1-(p-1)\sigma^{p-1}t]^{-\frac{1}{p-1}} & ((t,x) \in I_2) \\ 0 & ((t,x) \in I_0 \cup I_3)\ . \end{cases} \tag{3.11}$$

where the regions I_i, $i = 0, \ldots, 3$, are indicated in Fig.1.

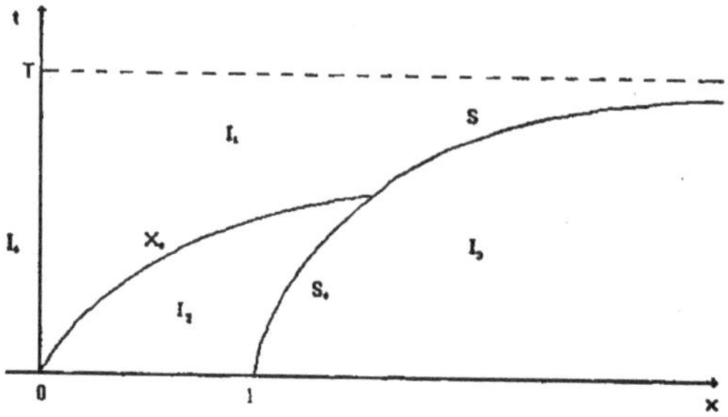

Fig.1

The function u is the unique entropy solution to the interaction problem (1.1) in Σ_T.

THEOREM 3.1. *Let $1 < p \leq m$, $u_0 = \sigma\chi_{[0,l]}$, $T > 0$ as given by Proposition 3.2 and u defined by (3.11). Then T is the blow-up time of u. Moreover:*

(i) there exist c_1, $c_2 > 0$ such that

$$c_1 \leq (T-t)^{\frac{1}{p-1}} \|u(t,\cdot)\|_{L^\infty} \leq c_2 \qquad \text{as } t \to T^-; \tag{3.12}$$

(ii) there exists $c > 0$ such that for any $x > 0$

$$\lim_{t\to T^-} x^{-\frac{1}{m-p}}[\sup_{y\in[0,x]} u(y,t)] \leq c \qquad \text{if } p < m, \tag{3.13}$$

$$\lim_{t\to T^-} e^{-x}[\sup_{y\in[0,x]} u(y,t)] \leq c \qquad \text{if } p = m. \tag{3.14}$$

It should be clear from the above result that in such case u can be continued beyond the blow-up time T in the sense of Definition 2.1. In fact, it suffices to make the following definition for $t \geq T$

$$u(t,x) := \begin{cases} t^{-\frac{1}{p-1}} f(x t^{\frac{m-p}{p-1}}) & x \geq 0 \\ 0 & x < 0 \,. \end{cases} \tag{3.15}$$

When $p > m$ the occurence and properties of blow-up depend on the quantity $\lambda := l\sigma^{p-m}$. Here we will describe only the case when (3.7) is satisfied, the other case being similar (see [NT1]).

Let T be given by Proposition 3.3. If the curves s and x_{BU} don't intersect, we set $T = +\infty$. Again the solution is given by (3.11). The regions I_i $(i = 0, \ldots, 3)$ are depicted in Fig.2 for the case $m < p \leq m+1$ and, if $p > m+1$, for $\lambda > M_1$ (see Proposition 3.3).

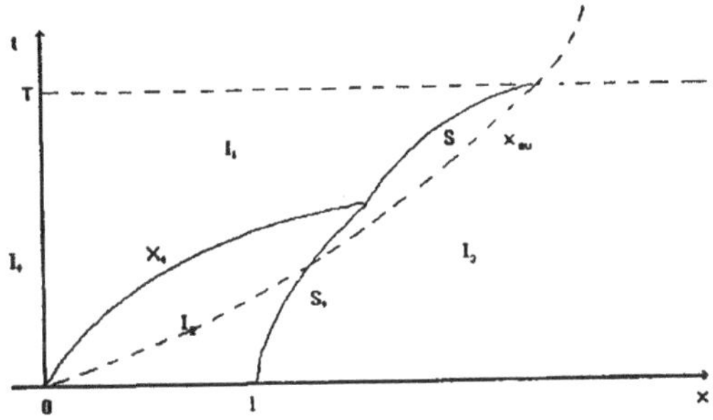

Fig.2

The same regions are given in Fig.3 if $p > m+1$ and $0 < \lambda \leq M_1$.

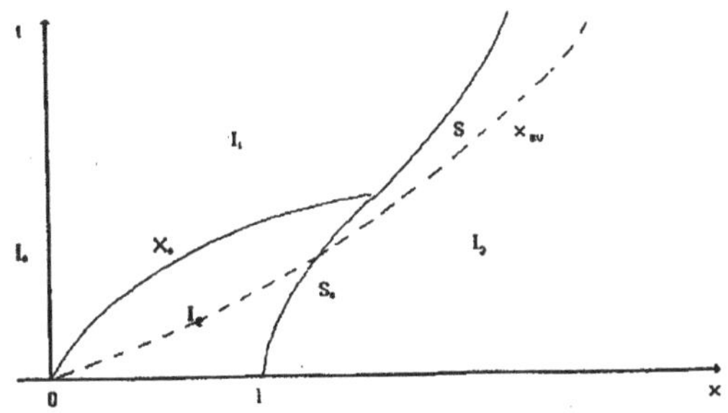

Fig.3

THEOREM 3.2. (i) *Let either $m < p \leq m+1$ or $p > m+1$ and $\lambda \in (M_1; M_0)$, where M_0 and M_1 are given in Proposition 3.3. Then T is the blow-up time of the solution u; moreover, there exist c_1, $c_2 > 0$ such that*

$$c_1 \leq \log(\frac{T}{t})^{\frac{1}{p-1}} \|u(t,\cdot)\|_{L^\infty} \leq c_2 \qquad \text{as } t \to T^-. \tag{3.16}$$

(ii) *Let $p > m+1$ and $\lambda \in (0, M_1)$. Then the solution u is global; moreover, there exist c_1, $c_2 > 0$ such that*

$$c_1 \leq t^{\frac{1}{m}} \|u(t,\cdot)\|_{L^\infty} \leq c_2 \qquad \text{as } t \to +\infty. \tag{3.17}$$

It should be clear from the above discussion that when blow-up occurs, even for $p > m$ the solution u can be continued by (3.16) beyond the blow-up time T in the sense of Definition 2.1. This is feasible in the restricted region

$$K := \{(x,t) | \, t \geq T, \;\; x < x_{BU}(t)\}.$$

REFERENCES

[BC] P. BARAS, L. COHEN, *Complete Blow-up after T_{max} for the Solution of a Semilinear Heat Equation*, J. Funct. Anal. **71** (1987), 142–174.

[Kr] S.N. KRUŽKOV, *First Order Quasilinear Equations in Several Independent Variables*, Mat. Sb. **81** (1970), 228–255. English transl.: Math. USSR Sb. **10** (1970), 217–243.

[La] P.D. LAX *Hyperbolic Systems of Conservation Laws and the Mathematical Theory of Shock Waves*, Conf. Board Math. Sci. **11**, SIAM (1973).

[Na] R. NATALINI, *Unbounded Solutions for Conservation Laws with Source*, Nonlin. Anal. TMA, to appear.

[NT1] R. NATALINI, A. TESEI, *Blow-up of Solutions for a Class of Balance Laws*, Quaderno IAC **1** (1993).

[NT2] R. NATALINI, A. TESEI, in preparation.

[Ol] O. OLEĬNIK *Discontinuous Solutions of Nonlinear Differential Equations*, Uspehi Mat. Nauk. **12** (1957), 3–73. English transl.: Amer. Math. Soc. Transl. **26** (1963), 95–172.

[SGKM] A.A. SAMARSKII, V.A. GALAKTIONOV, S.P. KURDYUMOV, A.P. MIKHAILOV, *Blow-up in Problems for Quasilinear Parabolic Equations*, Nauka, Moscow.

[Ug] M. UGHI, *On the Porous Media Equation with either Source or Absorption*, Cuadernos Inst. Mat. "Beppo Levi" **22** (1992).

SYMPLECTIC INTEGRATION FOR MACROMOLECULAR DYNAMICS

ROBERT D. SKEEL

JEFFREY J. BIESIADECKI

and

DANIEL OKUNBOR

Department of Computer Science, University of Illinois at Urbana-Champaign 1304 West Springfield Avenue, Urbana, Illinois 61801-2987, U.S.A.

ABSTRACT

Hamiltonian dynamical systems arise in applications such as astronomy, plasma physics, and molecular dynamics (MD), and, in particular, structural biology and drug design. Such systems have numerous special properties, some of which can be preserved by numerical integrators. In particular, methods that are *symplectic* preserve the Poincaré invariants and seem to give particularly good results, but the role of symplecticness is not well understood. We describe the results of numerical experiments with simulated liquid argon to test the importance of being symplectic. We discover that nonsymplectic methods can duplicate the good results of symplectic methods if velocity rescaling is used to conserve energy. For MD simulations only low accuracy is desired; nonetheless, it is possible that higher order symplectic integrators might be advantageous and we report on such experiments. In practical MD simulations with an underlying method that is symplectic a number of shortcuts are possible that compromise the symplectic property of the integrator—symplectic crimes. We enumerate and discuss these. In particular it has been found advantageous, because of the wide range of time-scales present, to use *multiple time steps*, in which different terms of the right-hand side are integrated with different time steps. Multiple time stepping is applicable to a wide range of practical problems. Most ways of doing multiple time steps convert a symplectic method into one that is not. It is shown how to use multiple time steps so as to preserve the symplectic property.

Recently, in a July 1992 issue of *Science* appeared a news story[1] concerning computational evidence for pervasive chaos in the solar system. Greater detail about the computation is given in a companion research article,[2] where it is mentioned that the numerical integration was performed by symplectic techniques. Evidently interest continues to grow in symplectic integrations, whose first published accounts[3,4,5] are still quite recent.

If there is one idea in this paper that is dominant, it concerns the structure of a system of ordinary differential equations (ODEs)

$$y' = f(y)$$

from a computation point of view.

Among specialists in the numerical solution of ODEs a common idea is that a system might be partitioned into subsystems

$$\begin{aligned}
\frac{d}{dt}y_1 &= f_1(y_1, y_2, \ldots, y_m), \\
\frac{d}{dt}y_2 &= f_2(y_1, y_2, \ldots, y_m), \\
&\vdots \\
\frac{d}{dt}y_m &= f_m(y_1, y_2, \ldots, y_m).
\end{aligned}$$

Computational advantages might be

- the use of *multi-rate* methods, in which different subsystems are integrated with different time steps,
- parallelism,
- acceleration of iterations.

This idea of partitioning finds notable use in waveform relaxation.

Perhaps a better idea, and one that is often used for PDEs, is to express the right-hand side as a sum of interactions:

$$\frac{d}{dt}y = \sum_{j=1}^{M} \phi_j(y)$$

where the $\phi_j(y)$ are vectors whose computation is essentially that of a scalar function. This may not always be possible but it is in a number of applications. Examples include a force and its reaction, a chemical reaction, a finite element (note that $\iint_\Omega = \iint_{\Omega_1} + \iint_{\Omega_2} + \cdots + \iint_{\Omega_M}$), a finite difference flux term, and an electric circuit element (possibly). The motivation for doing this may be to permit

- the use of *multiple time steps*, in which different terms of the right-hand side are integrated with different time steps. (However, it is true that by a change of variables multiple time-stepping can often be expressed as a multi-rate method.)
- parallelism,
- possible acceleration of iterations.

The paper is organized as follows:

1. molecular dynamics,
2. meaning of symplectic,
3. importance of being symplectic,
4. advantage of higher order,
5. symplectic crimes,
6. multiple time stepping,
7. constraint dynamics.

1. Molecular Dynamics

Applications of Hamiltonian systems include structural biology, pharmaceutical design, semiconductor modeling, superconductor modeling, plasma modeling, astrophysics, materials, and partial differential equations. The first five of these have been designated as "Grand Challenges" by the U.S. scientific establishment.

Molecular dynamics (MD) for large molecules is modeled by Newton's law of motion

$$m_i \frac{d^2}{dt^2}\vec{r}_i = -\frac{\partial}{\partial \vec{r}_i} V(\vec{r}_1, \vec{r}_2, \ldots, \vec{r}_N) \tag{1}$$

where m_i is the mass of the i-th atom, $\vec{r}_i$ is the position of the i-th atom, and V is the potential energy. The potential energy is a sum of contributions from long-range electrostatic interactions that fall off as $1/r$, short-range van der Waals interactions that fall off as $1/r^6$, and bonded interactions (covalent and hydrogen bonding):

$$V = \sum_{j>i}\sum V_{ij}(\|\vec{r}_j - \vec{r}_i\|) + \sum_{\text{bonds \& H bonds}} \cdot$$

The double summation over atom pairs consists of the electrostatic and van der Waals potentials. By far the major part of the computation is taken up by these nonbonded 2-body interactions, particularly, the long-range electrostatic force, which involves $N^2/2$ interactions. Also the electrostatic interactions cause the Jacobian matrix of the right-hand side to be dense, making fully implicit methods with Newton iterations impractical under normal circumstances. On the other hand it seems practical to use methods that employ analytical higher derivatives. In fact the use of higher derivatives might amortize the cost of doing square roots, which on some computer systems constitute half of the CPU time of an MD simulation.

Typical biomolecular simulations involve 2 to 20 thousand atoms. By contrast a virus has on the order of 1 million atoms. A typical integration interval is 100 picoseconds with a stepsize $\approx$ 0.001 psec.

It is notable that the ODEs are unstable, or chaotic. Numerical experiments[6] suggest a Lyapunov exponent that can be described as follows: perturbations grow 10-fold every 0.6 psec. This means we lose one digit after every 0.6 psec simulation time, and after 10 psec we have lost all accuracy of a double-precision simulation. So for perhaps 90% of the integration the simulation has no accuracy at all. You could say that one might as well be generating random numbers. And they do: as an alternative, researchers often use *stochastic dynamics*. This approach employs stochastic differential equations in an attempt to capture the long-time dynamics without resolving the chaotic thermal motion.

In addition to symplectic integration another important development of the 1980s that ought to be mentioned, even though it is somewhat of a digression, is the *Greengard-Rokhlin Fast Multipole Method* (FMM).[7] This method reduces the cost of calculating the electrostatic interaction from $O(N^2)$ to $O(N)$. It is based on a

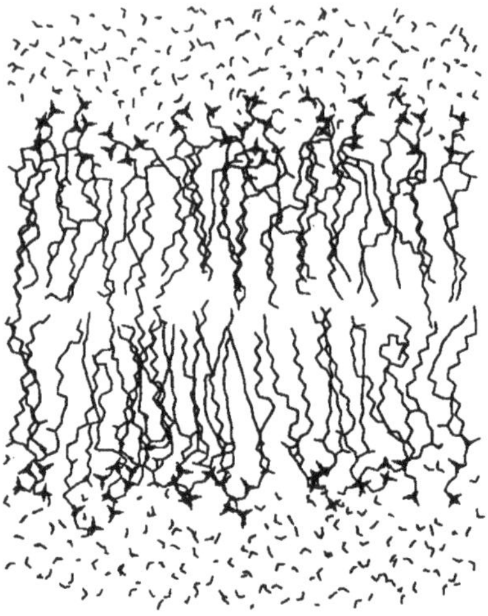

Figure 1: Bilipid membrane patch in water (courtesy of H. Heller)

hierarchical representation of the electrostatic interactions. Particle fields are represented at more distance locations by a hierarchy of multipole expansions. These are transferred to distant locations as Taylor expansions:

```
                    ------->
                   /        \
                  /          \
                 ------------->
                /              \
               /                \
multipole expansions---->Taylor expansions
             /                    \
            /                      \
particle fields---------->forces on particles
```

One of the best implementations, specifically for MD, is due to Board.[8] The accuracy of this method depends on the number of terms used in the expansions. With 4 terms the error is 30 ppm (parts per million) and with 8 terms the error is 3 ppm, which is nearly single-precision accuracy. If we express the cost of the direct method as

$$\text{constant} \cdot n^2,$$

then the FMM has a cost

$$3000 \cdot \text{constant} \cdot n.$$

(This is based on discussions with J. Board and with L. Greengard.)

2. Meaning of Symplectic

What it means for a numerical method to be symplectic is discussed in various papers, e.g., Sanz-Serna.[9] Here we provide a brief definition. For a Hamiltonian system

$$\frac{d}{dt}q = \frac{\partial}{\partial p}H(q,p), \qquad \frac{d}{dt}p = -\frac{\partial}{\partial q}H(q,p),$$

we can think of one step of a (one-step) numerical method as a mapping

$$\begin{pmatrix} q_n \\ p_n \end{pmatrix} \mapsto \begin{pmatrix} q_{n+1} \\ p_{n+1} \end{pmatrix}.$$

The analytical counterpart of this mapping is known as the Δt-flow of the system; and if it possesses a certain abstract property, it is said to be symplectic, or canonical. One consequence of this property, known as the *Liouville* property, is that any volume in (q,p) phase space remains unchanged by the Δt-flow. A straightforward test for the symplectic property in a numerical method is to verify that

$$\begin{pmatrix} \partial q_{n+1}/\partial q_n & \partial q_{n+1}/\partial p_n \\ \partial p_{n+1}/\partial q_n & \partial p_{n+1}/\partial p_n \end{pmatrix}^{\mathrm{T}} \begin{pmatrix} 0 & I \\ -I & 0 \end{pmatrix} \begin{pmatrix} \partial q_{n+1}/\partial q_n & \partial q_{n+1}/\partial p_n \\ \partial p_{n+1}/\partial q_n & \partial p_{n+1}/\partial p_n \end{pmatrix} = \begin{pmatrix} 0 & I \\ -I & 0 \end{pmatrix}.$$

The possibility of symplectic numerical integrators was discovered independently by Ruth[3] and Feng,[4] who published the first papers on this topic.

3. Importance of Being Symplectic

There is increasing empirical evidence favoring symplectic integrators, but the role of symplecticness is not well understood. In practice we compute a finite number of trajectories on finite time intervals, so the symplectic property does not strictly apply. Possibly it is some consequence of being symplectic that is important. Perhaps the best explanation of why it is important to use symplectic methods for Hamiltonian systems is that given by Sanz-Serna,[9] viz., that a symplectic method produces a numerical solution that is very nearly the exact solution of a nearby *Hamiltonian* system. This is particularly appropriate for MD, where the potential energy functions are empirical: their form is ad hoc and their parameters are fitted to quantum mechanics computations and experimental observables. Nonetheless even if we could justify the accuracy by such a backward error analysis argument, it is essential to know that the quantities that we wish to calculate are not sensitive to small errors in the Hamiltonian. A backward error analysis without a sensitivity analysis is of little value.

We have performed numerical experiments testing the importance of being symplectic, including some with simulated liquid argon.[10] Figure 2 shows for various

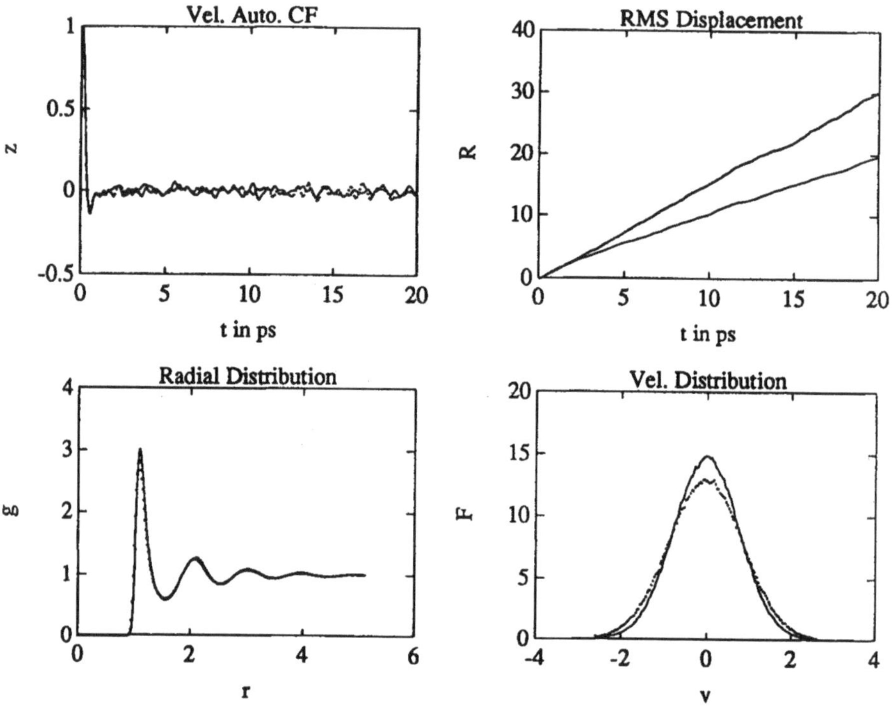

Figure 2: Various statistics for a symplectic and a nonsymplectic method

statistics of the simulation the difference between using a symplectic and a nonsymplectic method, both of order 4 and with time step $\Delta t = 0.0121$ femtoseconds. The dramatically different (and inferior) behavior of the nonsymplectic method can be explained, it happens, by the fact that it loses energy as time increases.

The poor energy conservation of most nonsymplectic methods motivated us to experiment with *velocity rescaling*—after every step uniformly scale velocities to conserve energy exactly. We discovered that this leads to results that are as good as those obtained with symplectic methods. Also we have obtained equally good results with *Liouville* methods, which merely conserve volume in phase space. Suris[11] showed for methods of order of accuracy at least 2 that the Liouville property is equivalent to the existence of stability interval for

$$\frac{d^2}{dt^2}y = -\omega^2 y.$$

4. Advantage of Higher Order

Currently both symplectic and nonsymplectic methods are used in molecular dynamics. Most popular is the Encke/Stőrmer/leapfrog/Verlet method, which happens to be symplectic but is only second-order accurate. It is natural to ask whether newly discovered methods of higher order might be better. The aim of MD simulations is not to reproduce the detailed motion of atoms but to calculate averages and macroscopic features of the dynamics. Hence only low accuracy is desired. Nonetheless, it is possible that higher order symplectic integrators might be advantageous. A typical stepsize for the Verlet method[12] applied to liquid argon is $\Delta t = 10$ fsec. To get the same accuracy as measured by RMS energy deviation, the fourth-order method of Calvo and Sanz-Serna[13] needs a scaled stepsize of only 14 fsecs, thus providing a 40% improvement in performance. (The scaled stepsize is the stepsize divided by the number of new force evaluations per step.)

5. Symplectic Crimes

In practical MD simulations with an underlying method that is symplectic, a number of shortcuts are possible that compromise the symplectic property of the integrator—symplectic crimes:

- finite precision. However, there is a very interesting result of Scovel[14] showing essentially that with fixed-point arithmetic we get the symplectic integration of a slightly different Hamiltonian system.
- incomplete iterations, due either to nonlinear constraints that must be satisfied or to the use of implicit integration schemes. It is common to impose constraints on bondlengths in order to remove the highest frequencies from the molecular structure.

- fast electrostatic solvers. There is the temptation to truncate the expansion before reaching the level of machine precision.
- cutoffs for nonbonded interactions. It is typical not to calculate electrostatic forces beyond some cutoff distance.
- multiple-time step techniques, which have been found advantageous, because of the wide range of time-scales present.
- variable time steps Δt_n. The varying of step size for an otherwise symplectic method seems to destroy the symplectic property.[15] This is ironic because a common argument for varying Δt is a Hamiltonian system, viz., a three-body problem in celestial mechanics.

6. Multiple Time Stepping

Multiple time steps was proposed in 1978,[16] for molecular dynamics. Let us begin by looking at a basic integration scheme without this feature. We can write our system of differential equations Eq. (1) as

$$\frac{d^2}{dt^2}R = M^{-1}F(R).$$

Given $R_n \approx R(n\Delta t)$, $\dot{R}_n \approx \frac{d}{dt}R(n\Delta t)$, $F_n = F(R_n)$, one step of (the velocity form of) the Verlet method is defined by

$$\begin{aligned} \dot{R}_{n+1/2} &= \dot{R}_n + \frac{\Delta t}{2}M^{-1}F_n, \\ R_{n+1} &= R_n + \Delta t \dot{R}_{n+1/2}, \\ F_{n+1} &= F(R_{n+1}), \\ \dot{R}_{n+1} &= \dot{R}_{n+1/2} + \frac{\Delta t}{2}M^{-1}F_{n+1}. \end{aligned}$$

We now give an example of multiple time step methods, also known as "distance class" methods.[17] Suppose we have three masses, two of them joined by a hard spring and another pair joined by a soft spring. We can express the potential energy as

$$V = V^{\mathrm{hard}} + V^{\mathrm{soft}}$$

and likewise the forces as

$$F = -\begin{bmatrix} \partial V/\partial \vec{r}_1 \\ \partial V/\partial \vec{r}_2 \\ \partial V/\partial \vec{r}_3 \end{bmatrix} = F^{\mathrm{hard}} + F^{\mathrm{soft}}.$$

For the regular Verlet method we simply have

$$F_n = F_n^{\mathrm{hard}} + F_n^{\mathrm{soft}}.$$

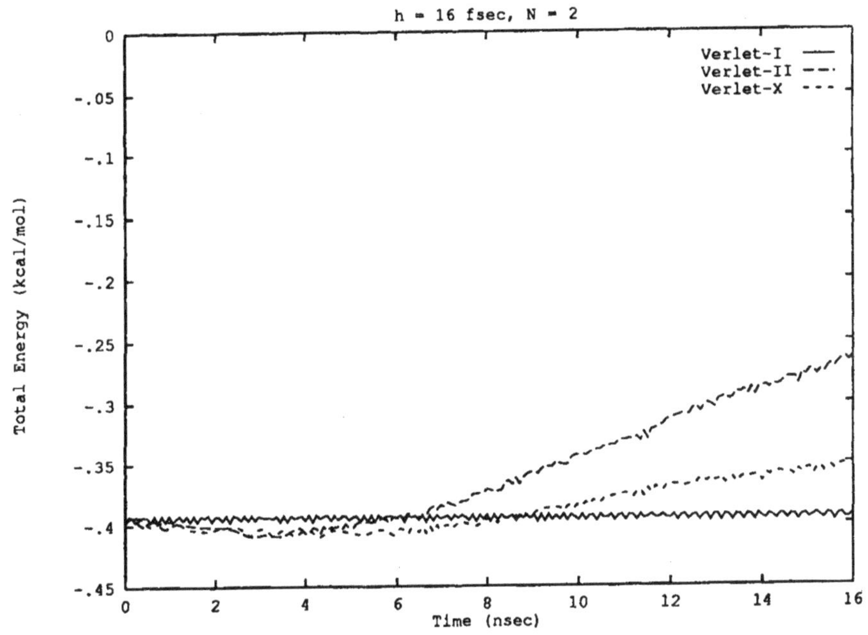

Figure 3: Energy conservation for various multiple time step schemes

A symplectic distance class scheme (Verlet-I method[17,18]) might have

$$F_n = \begin{cases} F_n^{\text{hard}} + 2F_n^{\text{soft}}, & n \text{ even}, \\ F_n^{\text{hard}}, & n \text{ odd}, \end{cases}$$

and thus avoid the computation of F_n^{soft} for odd values of n. A nonsymplectic distance class scheme (Verlet-X method[18]) might have

$$F_n = \begin{cases} F_n^{\text{hard}} + \frac{3}{2}F_n^{\text{soft}} - \frac{1}{2}F_{n-2}^{\text{soft}}, & n \text{ even}, \\ F_n^{\text{hard}} + F_{n-1}^{\text{soft}}, & n \text{ odd}, \end{cases}$$

and also avoid the computation of F_n^{soft} for odd values of n.

We[18] have performed numerous experiments comparing the symplectic distance class scheme to a couple of nonsymplectic schemes for both linear and nonlinear forces, and the outcome is always that the symplectic method is superior for a long enough integration. As an example we show in Figure 3 a one-dimensional simulation of three argon atoms in which neighboring atoms interact with Lennard-Jones potentials. For test purposes one of the time steps is double the other. Our parameters are the same

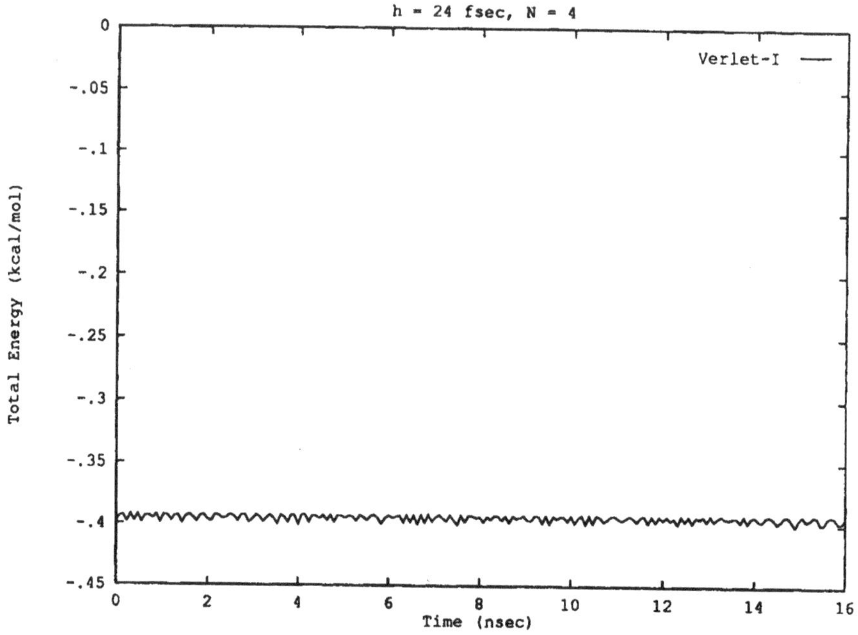

Figure 4: Energy conservation for the canonical multiple time step scheme

as those of Rahman[19] and initially all three particles start at rest with positions −3.6 Å, 0, and 3.6 Å. The plot shows that the symplectic scheme conserves energy very well for one million time steps but that the two nonsymplectic methods eventually become unstable. Figure 4 demonstrates the robustness of the symplectic method: the energy is conserved over the same span of simulated time even though the basic time step $\Delta t = h$ is increased and the ratio of the time steps is four. A typical time step for this problem is 10 fsec.

A drawback to these multiple time stepping schemes, but especially the symplectic one is that artificial "resonance" occurs[17,18] if the macro stepsize $N\Delta t$ has a frequency approximately equal to either the natural frequency or twice the natural frequency of the simulated system.

7. Constraint Dynamics

Constraints on the bondlengths, mentioned earlier, lead to a constrained system of

ODEs:

$$\frac{d^2}{dt^2}R = M^{-1}F(R), \quad g(R) = 0.$$

As this stands, it is overdetermined and we must introduce Lagrange multipliers λ to account for the forces that maintain the constraints. If we also convert the system to twice as many first-order equations, we get

$$\begin{aligned} \frac{d}{dt}R &= M^{-1}P, \\ \frac{d}{dt}P &= F(R) + g'(R)^{\mathrm{T}}\lambda, \\ g(R) &= 0, \end{aligned}$$

which is an index 3 differential-algebraic equation.[20]

As a consequence of the constraints on R there an equal number of constraints on P:

$$g'(R)M^{-1}P = 0.$$

We might ask whether we have a Hamiltonian system on the manifold $g(R) = 0$, $g'(R)M^{-1}P = 0$? The answer is yes.[21,22]

Symplectic methods for this problem were first proposed by Leimkuhler and Reich.[22] Recently, we[23] have shown that the popular method of imposing constraints, SHAKE,[24] is equivalent to another method RATTLE,[25] which we show to be both symplectic and time reversible. The SHAKE-like constraints are imposed thus:

$$\begin{aligned} R_{n+1} &= R_n + \Delta t M^{-1} P_{n+1/2}, \\ P_{n+1/2} &= P_{n-1/2} + \Delta t(F(R_n) + g'(R_n)^{\mathrm{T}}\lambda_n), \\ g(R_{n+1}) &= 0. \end{aligned}$$

By eliminating R_{n+1} and $P_{n+1/2}$, we can solve the last of these equations for λ_n by some iteration.

8. Acknowledgement

This work was supported in part by National Science Foundation Grant DMS 90 15533.

9. References

1. R.A. Kerr, From Mercury to Pluto, chaos pervades the solar system, *Science* **257** (1992) 33.

2. G.J. Sussman and J. Wisdom, Chaotic evolution of the solar system, *Science* **257** (1992) 56–62.

3. R.D. Ruth, A canonical integration technique, *IEEE Trans. Nuc. Sci.* **30** (1983) 2669–2671.

4. K. Feng, On difference schemes and symplectic geometry, Proceedings of the 1984 Beijing Symposium on Differential Geometry and Differential Equations—*Computation of Differential Equations*, ed. K. Feng, (Science Press, Beijing, 1985) 42–58.

5. P.J. Channell and J.C. Scovel, Symplectic integration of Hamiltonian Systems, *Nonlinearity* **3** (1990) 231–259.

6. M.P. Allen and D.J. Tildesley, *Computer Simulation of Liquids*, (Oxford University Press, 1987).

7. L. Greengard and V. Rokhlin, A fast algorithm for particle simulations, *J. Comput. Phys.* **73** (1987) 325–348.

8. J. A. Board, Jr., R. R. Batchelor, and J. F. Leathrum, Jr., in *Proceedings of the 5th AIAA/ASME Thermophysics and Heat Transfer Conference*, (1990) 27–34.

9. J.M. Sanz-Serna, Symplectic integrators for Hamiltonian problems: An overview, *Acta Numerica* **1** (1992) 243–286.

10. D. Okunbor, *Canonical Numerical Integrators for Hamiltonian Dynamical Systems*, Tech. Rept. UIUCDCS-R-92-1785, (Dept. Computer Sci., Univ. Illinois, Urbana, 1992); also, Ph.D. thesis.

11. Y.B. Suris, Some properties of methods for the numerical integration of systems of the form $\ddot{x} = f(x)$, *Zh. Vychisl. Mat. i Mat. Fiz.* **27** (1987) 1504–1515 (in Russian); same as *U.S.S.R. Comput. Maths. Math. Phys.* **27** (1987) 149–156.

12. L. Verlet, Computer 'experiments' on classical fluids. I. thermodynamical properties of Lennard-Jones molecules, *Phys. Rev.,* **159** (1967) 98–103.

13. M.P. Calvo and J.M. Sanz-Serna, The development of variable-step symplectic integrators, with applications to the two-body problem, *SIAM J. Sci. Statist. Comp.*, to appear.

14. C. Scovel, On symplectic lattice maps, *Physics Letters A,* **159** (1991) 396–400.

15. R.D. Skeel and C.W. Gear, Does variable step size ruin a symplectic integrator?, *Physica D* (1992), to appear.

16. W.B. Streett, D.J. Tildesley, and G. Saville, Multiple time step methods in molecular dynamics, *Mol. Phys.* **35** (1978) 639–648.

17. H. Grubmüller, H. Heller, A. Windemuth, and K. Schulten, Generalized Verlet algorithm for efficient molecular dynamics simulations with long-range interactions, *Molecular Simulation* **6** (1991) 121–142.

18. J. Biesiadecki and R.D. Skeel, Dangers of Multiple Time Step Methods, in preparation; also, J. Biesiadecki, Dangers of Using Multiple Time Step Algorithms for Molecular Dynamics Simulations, M.S. thesis, (Dept. Computer Sci., Univ. Illinois, 1993).

19. A. Rahman, "Correlations in the motion of atoms in liquid argon," *Phys. Rev. A, 136*, 405–411, 1964.

20. K.E. Brenan, S.L. Campbell, and L.R. Petzold, *Numerical Solution of Initial-Value Problems in Differential-Algebraic Equations*, (North-Holland, 1989).

21. R. Abraham and J.E. Marsden, *Foundations of Mechanics*, 2nd edition, (Benjamin Cummins Publishing Co., Reading, Mass., 1978).

22. B. Leimkuhler and S. Reich, *The Numerical Solution of Constrained Hamiltonian Systems*, Tech. Rept. SC-92-16, (Konrad-Zuse-Zentrum für Informationstechnik Berlin, Berlin, West Germany, 1992), submitted.

23. B. Leimkuhler and R.D. Skeel, Symplectic numerical integrators for Hamiltonian systems, in preparation.

24. J.P. Ryckaert, G. Ciccotti, and H.J.C. Berendsen, Numerical integration of the cartesian equations of motion of a system with constraints: molecular dynamics of n-alkanes, *J. Comput. Phys.* **23** (1977) 327–341.

25. H.C. Andersen, Rattle: a 'velocity' version of the shake algorithm for molecular dynamics calculations, *J. Comput. Phys.* **52** (1983) 24–34.

COMPLETELY DISCRETE METHODS FOR PARABOLIC EQUATIONS WITH MEMORY

V. Thomée
Department of mathematics, Chalmers University of Technology, S-412 96 Göteborg, Sweden

ABSTRACT
We survey recent work on numerical solution of parabolic integro-differential equations by finite elements in space, combined with standard time-stepping methods. Special attention is paid the approximation of the memory term by quadrature rules which lead to stability and limited storage requirements.

0. Introduction.

Our purpose in this paper is to present some recent work on Galerkin finite element methods for the numerical solution of initial-boundary value problems of the form

$$u_t + Au = \int_0^t B(t,s)u(s)ds + f(t), \quad \text{for } t \in J = (0,T], \tag{0.1}$$
$$u(0) = v.$$

Here $u = u(x,t)$ and $f = f(x,t)$ are real-valued functions on $\bar{\Omega} \times \bar{J}$, where Ω is a domain in R^d. Further, $u_t = \partial u/\partial t, A$ is a time-independent selfadjoint second order linear elliptic differential operator in Ω under homogeneous Dirichlet boundary conditions, with a positive smallest eigenvalue, and $B(t,s)$ is an arbitrary linear partial differential operator of order $\beta \leq 2$. The function f and the coefficients of A and $B(t,s)$ are assumed smooth in $\bar{\Omega} \times \bar{J}$, and we also assume that $\partial\Omega$ is smooth and that, except in Section 5 below, T is finite; without special mention, several of the constants C below may depend on T.

Such problems, and nonlinear variants thereof, occur in several applications, such as in heat conduction in materials with memory, compression of poro-viscoelastic media, nuclear reactor dynamics, etc., and have been extensively investigated in the mathematical literature, see, e.g., the references quoted in Yanik and Fairweather [12] and Thomée [8], where also early references can be found concerning the numerical solution by means of finite differences. Recently finite element methods for (0.1) have been studied in Sloan and Thomée [6], Yanik and Fairweather [12], Le Roux and Thomée [3], Thomée and Zhang [10], Cannon and Lin [1], Lin, Thomée, and Wahlbin [4], Chen, Thomée, and Wahlbin [2], Zhang [13], Pani, Thomée, and Wahlbin [5], Thomée and Zhang [11], and Thomée and Wahlbin [9]. The following is a survey of results from these papers, with special emphasis on discretization in time.

The plan of the paper is the following: In Section 1, following [10] and [4], we discuss the discretization with respect to the space variables of an equation of the form (0.1) by means of a standard Galerkin finite element method, and optimal order error estimates are shown under the assumption that the exact solution of (0.1) is sufficiently smooth. Section 2 is devoted to the discretization in time and follows [13], where ideas from [6] and [3] are further developed. Here the emphasis in on the choice of quadrature rules for the integral term, which combine stability of the resulting scheme with accuracy and reasonable storage requirements. In Section 3 we consider, for the homogeneous equation, the situation when the initial data are nonsmooth, and present error estimates from [10] and [11] which are of optimal order for positive time, provided the finite element space is not of too high an order, for both semidiscrete and completely discrete cases. In Section 4 we describe work from [2] in the case $B(t,s) = b(t-s)B$ where b is a weakly singular scalar kernel. In Section 5, finally, we consider the long time behavior of the solutions of the semidiscrete and completely discrete approximations according to [9].

1. The spacially semidiscrete problem.

In this section we present results from [10] concerning the discretization in space of the initial boundary value problem (0.1).

We want to determine an approximate solution of (0.1) in a finite element space $S_h \subset H_0^1$ belonging to a family $\{S_h\}$ which we assume to have the approximation property that, for some integer $r \geq 2$,

$$\inf_{\chi \in S_h} \{\|v-\chi\| + h\|v-\chi\|_1\} \leq Ch^s\|v\|_s, \quad \text{for } 1 \leq s \leq r.$$

Here and below, we denote by $\|\cdot\|$ the standard norm in $L_2 = L_2(\Omega)$, $\|\cdot\|_r$ that in the Sobolev space $H^r = H^r(\Omega)$, and by H_0^1 the elements in H^1 which vanish on $\partial\Omega$. For the purpose stated we write our problem in weak form, and pose the semidiscrete problem to find $u_h : \bar{J} \to S_h$ such that

$$\begin{aligned}(u_{h,t},\chi) + A(u_h,\chi) &= \int_0^t B(t,s;u_h(s),\chi)ds + (f(t),\chi), \quad \forall \chi \in S_h,\ t \in J,\\ u_h(0) &= v_h,\end{aligned} \tag{1.1}$$

where v_h is a suitable approximation of v in S_h. Here $(\cdot,\cdot)$ is the inner product in L_2 and $A(\cdot,\cdot)$ and $B(t,s;\cdot,\cdot)$ are bilinear forms on $H_0^1 \times H_0^1$ associated with the operators A and $B(t,s)$.

It is easy to see that the semidiscrete problem (1.1) admits a unique solution, and one may then prove the following error estimate which is identical to that valid for the purely parabolic differential equation, that is, for $B(t,s) \equiv 0$ (cf., e.g., [7]).

Theorem 1.1. *Under the appropriate regularity assumptions on the exact solution u of (0.1) we have, for the solution u_h of (1.1),*

$$\|u_h(t) - u(t)\| \le \|v_h - v\| + Ch^r\{\|v\|_r + \int_0^t \|u_t\|_r ds\}, \quad \text{for } t \in J.$$

We recall that the standard proof of the corresponding result in the purely parabolic case relies on the introduction of the Ritz projection $R_h : H_0^1 \to S_h$, defined by

$$A(R_h u - u, \chi) = 0, \quad \forall \chi \in S_h. \tag{1.2}$$

One writes the error

$$e = u_h - u = (u_h - R_h u) + (R_h u - u) = \theta + \rho,$$

and proceeds to estimate θ and ρ separately, noting that $\theta \in S_h$. Here, by the well-known error estimate for the elliptic problem, one has

$$\|\rho(t)\| + h\|\rho(t)\|_1 \le Ch^r\|u(t)\|_r \le Ch^r\{\|v\|_r + \int_0^t \|u_t\|_r ds\}, \quad \text{for } t \in \bar{J},$$

and a similar estimate is valid for ρ_t. With this choice we find now that θ satisfies

$$(\theta_t, \chi) + A(\theta, \chi) = \int_0^t B(t,s;e(s),\chi)\, ds - (\rho_t, \chi). \tag{1.3}$$

The standard energy argument, choosing $\chi = \theta(t)$, will then introduce a term containing $\|\rho\|_1$ on the right hand side, which, however, is only of order $O(h^{r-1})$, and which will thus lead to a loss of accuracy of one order of h in the error estimate. In [10] a more refined argument is used to avoid this loss.

We shall sketch an alternative approach to that in [10], invented by Cannon and Lin [1] and further developed in [4]. It consists in introducing instead of R_h a projection $V_h = V_h(t) : C(\bar{J}, H_0^1(\Omega)) \to C(\bar{J}; S_h)$, referred to as the Ritz-Volterra projection, by

$$A((V_h u - u)(t), \chi) + \int_0^t B(t,s;(V_h u - u)(s), \chi)\, ds = 0, \quad \text{for } \chi \in S_h,\ t \in \bar{J}.$$

Note that V_h is not a projection of functions of the spatial variable onto S_h, but operates on functions in both space and time. Although the estimation of $V_h u - u$ requires an extra effort, the analysis of $u_h - V_h u$ now becomes simpler than that of $u_h - R_h u$. Writing this time

$$e = u_h - u = (u_h - V_h u) + (V_h u - u) = \theta + \rho,$$

one may show that, for $t \in \bar{J}, \rho$ satisfies the estimates

$$\|\rho(t)\| + h\|\rho(t)\|_1 \leq Ch^r\{\|u(t)\|_r + \int_0^t \|u(s)\|_r ds\},$$

and

$$\|\rho_t(t)\| + h\|\rho_t(t)\|_1 \leq Ch^r\{\|u(t)\|_r + \|u_t(t)\|_r + \int_0^t (\|u\|_r + \|u_t\|_r) ds\}. \quad (1.4)$$

Assuming this we can now complete the proof of Theorem 1.1. With our present definitions, (1.3) is replaced by

$$(\theta_t, \chi) + A(\theta, \chi) = \int_0^t B(t, s; \theta(s), \chi)\, ds - (\rho_t, \chi).$$

Setting $\chi = \theta(t)$ and integrating with respect to t, we have this time, with $c > 0$,

$$\|\theta(t)\|^2 + c\int_0^t \|\theta\|_1^2 ds \leq \|\theta(0)\|^2 + C\int_0^t \|\rho_t\|\, \|\theta\| ds + C\int_0^t \int_0^s \|\theta\|_1^2\, d\tau\, ds.$$

Gronwall's lemma and the estimate for ρ_t from (1.4) now show

$$\|\theta(t)\| \leq C\{\|\theta(0)\| + \int_0^t \|\rho_t\| ds\} \leq Ch^r\{\|v\|_r + \int_0^t \|u_t\|_r ds\},$$

which is the desired estimate. The proof is complete.

2. Time stepping.

In this section we shall present results from [13] concerning discretization in both space and time of parabolic integro-differential equations with smooth solutions.

We shall first discuss discretization of the semidiscrete problem (1.1) based on the backward Euler approximation of the time derivative and using various quadrature formulas for the integral term. Let thus k be the time step, $t_n = nk, U^n$ the approximation of $u(t_n), f^n = f(t_n)$, and $\bar{\partial}_t U^n = (U^n - U^{n-1})/k$. Let further

$$\sigma^n(\phi) = \sum_{j=0}^{n-1} \omega_{nj}\phi(t_j) \approx \int_0^{t_n} \phi(s)\,ds, \tag{2.1}$$

where $\{\omega_{nj}\}_{j=0}^{n-1}$ are quadrature weights. The backward Euler type scheme now has the form

$$(\bar{\partial}_t U^n, \chi) + A(U^n, \chi) = \sigma^n(B(t_n, \cdot; U, \chi)) + (f^n, \chi), \quad \text{for } \chi \in S_h,\ t_n \in J,$$
$$U^0 = v_h. \tag{2.2}$$

Note that for simplicity we have chosen to not include U^n in the quadrature term.

An obvious example for (2.1) is provided by the rectangle rule

$$\sigma^n(\phi) = k\sum_{j=0}^{n-1} \phi(t_j), \tag{2.3}$$

thus by choosing $\omega_{nj} = k$ for $0 \le j \le n-1$. In this case we have for the quadrature error

$$q^n(\phi) = \sigma^n(\phi) - \int_0^{t_n} \phi(s)\,ds = O(k),$$

for smooth ϕ, which matches the truncation error for the difference quotient in (2.2).

A practical difficulty with this method is that all the U^n need to be stored as we step forward in time because they all enter in the subsequent equations. With this choice the number of U^n which have to be kept in the memory is thus of order $O(k^{-1})$.

In order to reduce the storage requirement we shall now propose a quadrature rule which is more sparse than the above, but nevertheless retains the order of accurracy of the backward Euler discretization. To accomplish this we base the quadrature rule on the longer time step $k_1 = mk$, where $m = [k^{1/2}]$, the integral part of $k^{1/2}$, thus with $k_1 = O(k^{1/2})$. Setting $\bar{t}_j = jk_1, j = 0, 1, \cdots$, we introduce j_n as the largest integer such that $\bar{t}_{j_n} \le t_{n-1}$. We then apply the trapezoidal rule with mesh-size k_1 on $(0, \bar{t}_{j_n})$ and the rectangle rule with mesh-size k on the remaining part, i.e., we define

$$\sigma^n(\phi) = k_1\{\frac{1}{2}\phi(0) + \phi(\bar{t}_1) + \cdots + \frac{1}{2}\phi(\bar{t}_{j_n})\}$$
$$+ k\{\phi(\bar{t}_{j_n}) + \phi(\bar{t}_{j_n} + k) + \cdots + \phi(t_{n-1})\}, \tag{2.4}$$

and obtain for the quadrature error $q^n(\phi) = O(k_1^2) + O(k_1 k) = O(k)$, as $k \to 0$, provided ϕ is smooth enough. The number of U^n that need to be stored thus reduces to $O(k_1^{-1}) + O(m) = O(k^{-1/2})$, without loss of accuracy.

It is possible to reduce the storage requirement even further, to $O(k^{-1/4})$, by using Simpson's formula with a stepsize of order $O(k^{1/4})$ for the main part of the interval of integration, cf. [13].

We shall say that the quadrature formula σ^n in (2.1) has dominated weights if there are numbers $\omega_j, j = 0, 1, \cdots$, independent of n, such that

$$|\omega_{nj}| \le \omega_j, \quad \text{for } 0 \le j < n, \quad \text{with } \sum_{j=0}^{n} \omega_j \le C, \quad \text{for } t_n \in J. \tag{2.5}$$

The rectangle rule obviously has this property, with $\omega_j = k$, and so has the modified trapezoidal rule described above, with $\omega_j = Ck_1$, if $j \equiv 0 (\mathrm{mod}\ m)$, and $\omega_j = Ck$, otherwise, since then $\sum_{j=0}^{n} \omega_j \le C j_n k_1 + Cnk \le CT$. Also the above mentioned quadrature rule based on Simpson's formula has dominated weights.

This property of σ^n will permit us to show a stability property which is needed for the error analysis. In order to do so, we define the discrete analogues $A_h : S_h \to S_h$ and $B_h = B_h(t,s) : S_h \to S_h$ of A and $B = B(t,s)$ by

$$(A_h\psi, \chi) = A(\psi, \chi), \quad (B_h(t,s)\psi, \chi) = B(t, s; \psi, \chi), \quad \forall \psi, \chi \in S_h, \ 0 \le s \le t \in \bar{J},$$

We now say that B_h is dominated by A_h if $B_h A_h^{-1}$ and $A_h^{-1} B_h$ are bounded with respect to the L_2 norm, uniformly in h. This holds, as is easily verified, if B is an operator of the form αA + a first order operator, or if S_h satisfies a standard inverse assumption. We note that the corresponding property for the continuous problem follows from the elliptic regularity estimate

$$\|B\phi\| + \|B^*\phi\| \le C\|\phi\|_2 \le C\|A\phi\|, \quad \text{for } \phi \in H^2 \cap H_0^1.$$

Lemma 2.1. *Let $B_h(t,s)$ and $B_{h,t}(t,s)$ be dominated by A_h, and assume that the quadrature rule σ^n has dominated weights. Then we have for the solution of (2.2) that*

$$\|U^n\| \le C(\|v_h\| + k \sum_{j=1}^{n} \|f^j\|), \quad \text{for } t_n \in J.$$

For the proof one notes that (2.2) may be written, with P_h the L_2-projection onto S_h,

$$\bar{\partial}_t U^n + A_h U^n = \sigma^n(B_h(t_n, \cdot)U) + P_h f^n, \quad \text{for } n \ge 1, \quad U^0 = v_h,$$

whence, with $E_{kh} = (I + kA_h)^{-1}$ and $\sigma^n(B_h U) = \sigma^n(B_h(t_n, \cdot)U)$,

$$U^n = k\sum_{j=0}^{n-1} E_{kh}^{n-j}\sigma^{j+1}(B_h U) + k\sum_{j=0}^{n-1} E_{kh}^{n-j} P_h f^{j+1} = I + II.$$

Using our definitions and assumptions one may show that

$$\|I\| \leq C\sum_{j=0}^{n-1} \omega_j \|U^j\|.$$

Together with the well-known stability property of E_{kh}^n and the boundedness of P_h, this yields

$$\|U^n\| \leq k\sum_{j=1}^{n} \|f^j\| + C\sum_{j=0}^{n-1} \omega_j \|U^j\|.$$

The desired result therefore follows from the following unconventional discrete version of Gronwall's lemma.

Lemma 2.2. *Assume that the nonnegative sequence* $\{g_n\}_{n=0}^{\infty}$ *satisfies*

$$g_n \leq \gamma_n + \sum_{j=0}^{n-1} \omega_j g_j, \quad \text{for } n \geq 0,$$

where $\{\omega_n\}_0^{\infty}$ *and* $\{\gamma_n\}_0^{\infty}$ *are nonnegative and nondecreasing, respectively. Then*

$$g_n \leq \gamma_n \exp\Big(\sum_{j=0}^{n-1} \omega_j\Big), \quad \text{for } n \geq 1.$$

We shall now turn to error estimates, and begin with a preliminary such result, in which the choice of the quadrature rule is not yet specified. Instead, the error bound contains one term which depends on this choice. The operator R_h which appears in this term is the standard Ritz projection defined in (1.2). Note that $R_h\phi = A_h^{-1}A\phi$ for smooth ϕ.

Theorem 2.1. *Let* $B_h(t,s)$ *and* $B_{h,t}(t,s)$ *be dominated by* A_h, *and assume that the quadrature rule* σ^n *has dominated weights. Let* $\|v_h - v\| \leq Ch^r\|v\|_r$. *Then we have for the solutions of (2.2) and (0.1) that*

$$\|U^n - u(t_n)\| \leq Ch^r\{\|v\|_r + \int_0^{t_n} \|u_t\|_r ds\} + Ck\int_0^{t_n} \|u_{tt}\| ds + C\max_{1\leq j\leq n} \|Q_h^j(R_h u)\|,$$

where $Q_h^n(\phi) = k\sum_{i=0}^{n-1} E_{kh}^{n-i} q^{i+1}(B_h\phi)$, with $q^j(B_h\phi) = q^j(B_h(t_j,\cdot)\phi)$.

In order to arrive at a complete error estimate, we now have to bound the global quadrature error $Q_h^n(R_h u)$. By a straightforward use of the stability of E_{kh}^n we find that

$$\max_{1\le j\le n} \|Q_h^j(R_h u)\| \le C \max_{1\le j\le n} \|q^j(B_h R_h u)\|.$$

For the rectangle rule (2.3) we have at once, if B_h and $B_{h,t}$ are dominated by A_h, that

$$\|q^j(B_h R_h u)\| \le Ck\int_0^{t_j} \|\frac{d}{ds}(B_h(t_j,s)A_h^{-1}Au(s))\|ds \le Ck\int_0^{t_j} (\|u\|_2 + \|u_t\|_2)ds,$$

and hence,

$$\max_{1\le j\le n} \|Q_h^j(R_h u)\| \le Ck\int_0^{t_n} (\|u\|_2 + \|u_t\|_2)ds.$$

However, using the smoothing property of E_{kh}^n, one may reduce the regularity requirements in the space variables. For the two quadrature rules considered in the beginning of this section we will then have the following.

Theorem 2.2. *Let $B_h(t,s)$ and $B_{h,t}(t,s)$ be dominated by A_h and let $\|v_h - v\| \le Ch^r\|v\|_r$. Then we have, for the rectangle rule (2.3) and for the modified trapezoidal rule (2.4), when $t_n \in J$,*

$$\|U^n - u(t_n)\| \le Ch^r\{\|v\|_r + \int_0^{t_n} \|u_t\|_r)ds\} + Ck\{\|v\| + \int_0^{t_n} (\|u_t\| + \|u_{tt}\|)ds\}.$$

It is natural to attempt to construct a more accurate time stepping scheme by taking the Crank-Nicolson method as a basis, and thus approximating (1.1) by

$$(\bar\partial U^n,\chi) + A(\bar U^n,\chi) = \sigma^n(B(t_{n-1/2},\cdot;U,\chi)) + (f^{n-1/2},\chi), \quad \text{for } \chi \in S_h,\ t_n \in J,$$

or, equivalently,

$$\bar\partial U^n + A_h\bar U^n = \sigma^n(B_h(t_{n-1/2},\cdot)U) + P_h f^{n-1/2}, \quad \text{for } t_n \in J,$$

where $\bar U^n = (U^n + U^{n-1})/2$, $t_{n-1/2} = (n-1/2)k$, and $f^{n-1/2} = f(t_{n-1/2})$. This time, instead of (2.1), $\sigma^n(\phi)$ should approximate the integral with upper limit $t_{n-1/2}$.

In this case the discrete solution operator corresponding to the homogeneous purely parabolic equation is E_{kh}^n, where now $E_{kh} = (I + kA_h/2)^{-1}(I - kA_h/2)$.

Because this operator does not have the smoothing property of the backward Euler method, the assumption that the quadrature rule has dominated weights does not suffice for stability, but a further condition has to be added. We thus assume that the quadrature rule has persistent dominated weights $\{\omega_{ni}\}$, which we shall take to mean that they are dominated by $\{\omega_i\}$ and, in addition, that they satisfy

$$\sum_{j=i+1}^{n-1} |\omega_{j+1,i} - \omega_{j,i}| \leq C\omega_i, \quad \text{for } 0 \leq i < n-1, \ t_n \in J.$$

Under this assumption it is possible to show a preliminary error estimate analogous to Theorem 2.1, but which is second order in time.

Quadrature rules satisfying this condition and yielding a $O(h^r + k^2)$ error estimate may be constructed based on the trapezoidal rule with stepsize k and on Simpson's rule with stepsize $k_1 = O(k^{1/2})$, with storage requirements of orders $O(k^{-1})$ and $O(k^{-1/2})$, respectively.

A second order time discretization based on two-step backward differencing was studied in [5]. Here the analysis was carried out by the energy method, which permits also time dependet operators A. A further increase of the accuracy to $O(h^r + k^3)$ may be accomplished by combining a subdiagonal third order Padé approximation of the differential equation part of (1.1) with a carefully chosen quadrature formula. We refer to [13] for details.

3. Nonsmooth initial data.

The error estimates of Sections 1 and 2 above require a degree of regularity of the solution, which is sometimes not present in practice. In this section we shall therefore present results from [10] and [11] concerning the homogeneous equation, i.e., (0.1) with $f = 0$, in the case that the initial data are only in $L_2(\Omega)$.

We recall (cf. [7], Chapter 3) that for the purely parabolic homogeneous differential equation, i.e., when $B \equiv 0$ and $f \equiv 0$ in (0.1), and when u_h is the solution of the corresponding spatially discrete problem (1.1), with v_h chosen as $P_h v$, then

$$\|u_h(t) - u(t)\| \leq Ch^r t^{-r/2} \|v\|, \quad \text{for } t > 0,$$

which shows an optimal order error estimate for positive time even when the initial data are only in L_2. This is related to the fact that the solution $u(t) = E(t)v$ of the homogeneous parabolic equation itself is smooth for positive time, and that, with s an arbitrary positive number,

$$\|u(t)\|_s = \|E(t)v\|_s \leq Ct^{-s/2}\|v\|, \quad \text{for } t > 0. \tag{3.1}$$

The solution operator $E(t)$ in this case is an analytic semigroup and is generated by $-A$.

The smoothing property of (3.1) carries over to the integro-differential equation only to a restricted extent.

Theorem 3.1. *We have, for the solution of (0.1) with $f = 0$,*

$$\|u(t)\|_s \leq Ct^{-s/2}\|v\|, \quad \text{for } t \in J, \text{ when } s \leq 4 - \beta.$$

One may show that this result is sharp in the sense that, in general, if $\beta = 2$ and $s > 2$, an inequality of the form $\|u(t)\|_s \leq C(t)\|v\|$, for all $v \in L_2$, cannot hold for any $t > 0$.

In spite of the restricted smoothness of the solution in the space variable, the regularity with respect to time is unlimited: One may prove that, for the solution of (0.1) with $f = 0$,

$$\|D_t^j u(t)\|_s \leq Ct^{-j-s/2}\|v\|, \quad \text{for } t \in J, \ j \geq 0, \ 0 \leq s \leq 2, \quad \text{where } D_t = \partial/\partial t.$$

We refer to [11] for details.

We now turn to the semidiscrete problem. The limited regularity of the exact solution of (0.1) will reflect on the semidiscrete problem, and we are only able to prove an optimal order error estimate at positive time if $r \leq 4 - \beta$, where $\beta \leq 2$ is the order of $B(t,s)$.

Theorem 3.2. *We have, for the solutions of (0.1) and (1.1), with $f = 0$ and $v_h = P_h v$,*

$$\|u_h(t) - u(t)\| \leq Ch^\gamma t^{-\gamma/2}\|v\|, \quad \text{for } t \in J, \text{ where } \gamma = \min(4 - \beta, r). \tag{3.2}$$

We shall now very briefly discuss the discretization also in time, using the case $f = 0$ of (2.2). The purpose is to determine conditions on the quadrature rule σ^n which yield good error estimates for nonsmooth v. We recall that for the purely parabolic problem ($B(t,s) = f = 0$), we have for the error in the backward Euler time discretization of (1.1) that (cf. [7])

$$\|U^n - u_h(t_n)\| \leq Ckt_n^{-1}\|v\|, \quad \text{for } t_n > 0.$$

Because of the singularity at $t = 0$ of the solution of the integro-differential equation we need to assume that the quadrature formula is appropriate for this situation, and we shall therefore, more precisely, assume that

$$|q^n(\phi)| \leq Ck \log\frac{1}{k}, \quad \text{if } |D_t^j\phi(t)| \leq C_j t^{-j}, \quad \text{for } t \in J, \ j \geq 0. \tag{3.3}$$

We then have the following:

Theorem 3.3. *Assume that $B_h(t,s)$ and its derivatives are dominated by A_h. Assume that σ^n has dominated weights $\{\omega_j\}$ which satisfy*

$$\sum_{j=1}^{n-1} \omega_j t_j^{-1} \le C \log \frac{1}{k}, \quad \text{for } t_n \in J, \tag{3.4}$$

and further that (3.3) holds. Then we have

$$\|U^n - u_h(t_n)\| \le Ck\big(t_n^{-1} + (\log \frac{1}{k})^2\big)\|v\|, \quad \text{for } t_n \in J. \tag{3.5}$$

It can easily be seen that conditions (3.3) and (3.4) hold for the rectangle rule (2.3) described above.

Condition (3.4) also holds for the trapezoidal rule (2.4), but now the main contribution to the quadrature error turns out to be $Ck_1^2 \int_{k_1}^{t_n} |\phi_{tt}| ds$, which only gives an error bound of order $O(k_1) = O(k^{1/2})$ under the assumptions of (3.3), thus with a loss of accuracy.

We shall therefore present an alternative quadrature rule which handles nonsmooth data better. We introduce this time the sequence of time steps defined by $\bar{t}_j = j^2 k$, $j = 0, 1, \cdots$, and let j_n be the largest integer for which $\bar{t}_{j_n} \le t_{n-1}$. We now propose to use the basic trapezoidal rule on the intervals $(\bar{t}_j, \bar{t}_{j+1})$ and then the rectangle rule with time step k on $(\bar{t}_{j_n}, t_{n-1})$, i.e., we set

$$\sigma^n(\phi) = \frac{1}{2} \sum_{j=0}^{j_n - 1} (\bar{t}_{j+1} - \bar{t}_j)(\phi(\bar{t}_{j+1}) + \phi(\bar{t}_j)) + k \sum_{j=j_n^2}^{n-1} \phi(t_j), \quad \text{for } n \ge 1.$$

It is easy to see that the number of nonzero weights in $\sigma^n(\phi)$ is $n - j_n^2 + j_n \le 3j_n + 1$ and that thus, like the earlier modified trapezoidal rule (2.4), the number of levels of the solution which need to be stored is $O(k^{-1/2})$.

One may show that (3.3) and (3.4) and hence the error estimate (3.5) hold in this case also. In both cases the estimate (3.5) may be combined with (3.2) to obtain a complete error estimate.

4. Weakly singular kernel.

Following [2] we shall now make some brief remarks on the initial value problem (0.1) when the operator $B(t,s)$ has the form $b(t-s)B$ where B a general partial differential operator of second order with smooth, time independent coefficients, and where b is a weakly singular scalar kernel such that

$$|b(t)| \le Ct^{-\alpha}, \quad \text{with } 0 \le \alpha < 1, \quad \text{for } t \in J.$$

For the semidiscrete solution one may then show the error estimate

$$\|u_h(t) - u(t)\| \leq Ch^2\{\|v\|_2 + \int_0^t \|u_t\|_2 ds\}, \quad \text{for } t \in J.$$

We consider now a backward Euler type discretization of the form (2.2). Since the integrand is singular even when the solution is smooth, we use product integration to approximate the memory term: In $\int_0^{t_n} b(t_n - s)\phi(s)ds$ we approximate ϕ by the piecewise constant function taking the value $\phi(t_j)$ in (t_j, t_{j+1}), and thus use

$$\sigma^n(\phi) = \sum_{j=0}^{n-1} \int_{t_j}^{t_{j+1}} b(t_n - s)\phi(t_j)ds = \sum_{j=0}^{n-1} \beta_{n-j}\phi(t_j), \quad \text{where } \beta_j = \int_{t_{j-1}}^{t_j} b(s)ds.$$

For this completely discrete method one may show the following:

Theorem 4.1. *Under our present assumptions, we have*

$$\|U^n - u(t_n)\| \leq C(h^2 + k)\{\|v\|_2 + \int_0^{t_n} (\|u_{tt}\| + \|u_t\|_2)ds\}, \quad \text{for } t \in J.$$

5. Long time behavior.

In this section we shall quote some results from [9] concerning the long-time approximation of the solution to the initial-value problem (0.1) when $B(t,s) = b(t-s)B$, where B is again a partial differential operator of second order, and where now the scalar kernel b is smooth and exponentially decreasing. We also assume that the memory term is dominated by the elliptic term so that, for some K, $\alpha > 0$, we have

$$(i) \quad |b(t)| \leq Ke^{-\alpha t} \quad \text{and} \quad (ii) \quad \beta = \int_0^\infty |b(s)|ds < 1/c_0. \tag{5.1}$$

Here $c_0 > 0$ is a constant such that $|(Bv, w)| \leq c_0\|A^{1/2}v\|\|A^{1/2}w\|$.

In our earlier discussion the energy arguments depend on Gronwall's lemma, and, applied to (0.1), give a bound for $u(t)$ that grows exponentially with t. However, under our present assumptions one may show the long-time stability property

$$\|u(t)\| \leq e^{-\gamma t}\|v\| + \int_0^t e^{-\gamma(t-s)}\|f(s)\|ds, \quad \text{for } t \geq 0, \quad \text{with } \gamma > 0.$$

Our purpose is therefore now to try to derive an analogous long-time stability property for the discrete problem and an associated error estimate. It turns out that this is considerably easier for the case $B = A$ than in the general case, due to the possibility of then using separation of variables. In this regard we have the following:

Proposition 5.1. *Assume that there are positive* $\tilde{\beta}, \bar{\gamma}$, *and* $\bar{k}$, *such that, with* $b_\gamma(t) = e^{\gamma t} b(t)$, $b_{\gamma n}(s) = b_\gamma(t_n - s)$,

$$\tilde{\beta}_{\gamma n} \equiv \sigma^n(|b_{\gamma n}|) \le \tilde{\beta} < 1, \quad \text{for } \gamma \le \bar{\gamma},\ k \le \bar{k},\ n \ge 1. \tag{5.2}$$

Then there are positive constants k_0 *and* γ *such that, for the solution of (2.2) (with* $B(t,s) = b(t-s)A$*),*

$$\|U^n\| \le e^{-\gamma t_n}\|U^0\| + k\sum_{j=1}^{n} e^{-\gamma t_{n-j}}\|f^j\|, \quad \text{for } n \ge 0,\ k \le k_0,$$

For an error estimate we assume that the quadrature rule is such that, for some integer ν,

$$|q^n(\phi)| \equiv \left|\sigma^n(\phi) - \int_0^{t_n} \phi(s)ds\right| \le Ck^{j/\nu}\int_0^{t_n} |\phi(s)|_j ds, \quad \text{for } n \ge 1,\ j = 1, \nu,$$

where $|\phi(s)|_j = \sum_{i=0}^{j} |D_s^i \phi(s)|$. We further assume that $|b(t)|_\nu \le Ce^{-\alpha t}$, cf. (5.1i).

One may then show that (5.2) holds and prove the following:

Theorem 5.1. *Under our present assumptions, let* $v_h = R_h v$. *Then there exist positive* C, δ, *and* k_0 *such that*

$$\|U^n - u(t_n)\| \le Ch^r\Big(\|u(t_n)\|_r + \int_0^{t_n} e^{-\delta(t_n - s)}\|u_t\|_r ds\Big)$$
$$+ Ck\int_0^{t_n} e^{-\delta(t_n-s)}(\|u_{tt}(s)\| + \sum_{i=0}^{q}\|D_s^i u(s)\|_2)ds, \quad \text{for } k \le k_0,\ n \ge 0.$$

Again the rectangle rule (2.3) satisfies the assumptions. In order to save storage space one may also use a rule based on a trapezoidal rule similar to (2.4). In this case it is not necessary to subdivide the tail interval $[\bar{t}_j, t_{n-1}]$ because the condition that the quadrature weights be dominated is not needed. Instead the somewhat simpler rule

$$\sigma^n(\phi) = \frac{k_1}{2}\sum_{j=1}^{j_n}(\phi(\bar{t}_j) + \phi(\bar{t}_{j-1})) + \frac{1}{2}(t_{n-1} - \bar{t}_{j_n})(\phi(t_{n-1}) + \phi(\bar{t}_{j_n})) + k\phi(t_{n-1})$$

will be adequate. In addition to the backward Euler method a second order backward difference method was also discussed in [9] and a $O(h^r + k^2)$ error estimate analogous to the above was obtained under the appropriate assumptions.

In [9] the completely discrete solution was studied also in the case that the operator B is different from A. Our result in this more general situation may be formulated as follows: Assume that $|b(t)| \leq K$, that $\|A^{-1}B\| \leq c_1$, and further that the quadrature weights are dominated in the sense of (2.5). Then, with $\gamma = \frac{2}{5}\min(\alpha, \lambda_0)$, where λ_0 is the smallest eigenvalue of A and $\tilde{\omega}_{ni} = \sum_{s=i}^{n-1} \omega_s$,

$$\|U^n\| \leq e^{Kc_1\tilde{\omega}_{n0}-\gamma t_n}(\|U^0\| + k\sum_{j=1}^{n} e^{Kc_1\tilde{\omega}_{nj}-\gamma t_{n-j}}\|f^j\|), \quad \text{for } n \geq 0.$$

For the rectangle rule (2.3) and the trapezoidal rule (2.4) we have $\tilde{\omega}_{nj} \leq c_2 t_{n-j} + c_3$. Our estimate therefore shows long–time stability provided Kc_1 is sufficiently small compared to α and λ_0 (and provided we have appropriate bounds on $\|f^j\|$).

REFERENCES

1. J. R. Cannon and Y. Lin, Nonclassical H^1 projection and Galerkin methods for nonlinear parabolic integro–differential equation, *Calcolo* **25** (1988), 187–201.
2. C. Chen, V. Thomée, and L.B. Wahlbin, Finite element approximation of a parabolic integro-differential equation with a weakly singular kernel., *Math.Comp.* **58** (1992), 587-602.
3. M.-N. Le Roux and V. Thomée, Numerical solution of semilinear integrodifferential equations of parabolic type with nonsmooth data, *SIAM J. Numer. Anal. 26* (1989), 1291-1309.
4. Y. Lin, V. Thomée, and L. B. Wahlbin, Ritz-Volterra projections to finite element spaces and applications to integro-differential and related equations, *SIAM J. Numer. Anal.* **28** (1991), 1047-1070.
5. A.K. Pani, V. Thomée, and L.B. Wahlbin, Numerical methods for hyperbolic and parabolic integro-differential equations., *J. Integral Equations (to appear).*
6. I. H. Sloan and V. Thomée, Time discretization of an integro-differential equation of parabolic type, *SIAM J. Numer. Anal.* **23** (1986), 1052-1061.
7. V. Thomée, *Galerkin Finite Element Methods for Parabolic Problems*, vol. No. 1054, Springer Lecture Notes in Mathematics, 1984.
8. V. Thomée, On the numerical solution of integro-differential equations of parabolic type, *International Series of Numerical Mathematics* **86** (1988), 477-493.
9. V. Thomée and L. B. Wahlbin, Long time numerical solution of a parabolic equation with memory, *Preprint, Department of Mathematics, Chalmers University of Technology, Göteborg* **1992-14**.
10. V. Thomée and N.-Y. Zhang, Error estimates for semi-discrete finite element methods for parabolic integro-differential equations, *Math. Comp.* **53** (1989), 121-139.
11. V. Thomée and N.-Y. Zhang, Backward Euler type methods for parabolic integro-differential equations with nonsmooth data, *Under preparation.*
12. E. G. Yanik and G. Fairweather, Finite element methods for parabolic and hyperbolic integro–differential equations, *Nonlinear Anal., Theory, Methods and Appl.* **12** (1988), 785-809.
13. N.–Y. Zhang, On fully discrete Galerkin approximations for partial integro–differential equations of parabolic type, *Math. Comp.* (to appear).

A Finite Element Method for the Poisson Equation in a Planar Exterior Domain

USHIJIMA, Teruo *), AJIRO, Atsushi **) and YOKOMATSU, Daisaku *)

*) Department of Computer Science and Information Mathematics
The University of Electro-Communications
1-5-1 Chofugaoka, Chofu-shi, Tokyo, 182, Japan

**) SONY Corporation

Abstract

A finite element method applied to the Poisson equation in an exterior planar domain is considered. The inhomogeneous term f of the equation is assumed to have a compact support contained in a disc with a sufficiently large radius. The problem is transformed to an interior problem with a non-local boundary condition imposed on the boundary of the disc. Error estimates for finite element approximation problems are given together with the report of the numerical experience for a proposed scheme of the problem in this paper.

Introduction

A finite element method for the Poisson equation (E) in a planar exterior domain Ω is the subject of the paper:

$$(E)\quad \begin{cases} -\Delta u = f & \text{in } \Omega, \\ u = 0 & \text{on } \Gamma, \\ \sup\limits_{|x|>a} |u| < \infty, & \end{cases}$$

where Γ is the boundary of Ω, a is chosen so that the obstacle $\mathcal{O} = (\Omega \cup \Gamma)^C$ may be contained in the disc D_a with the radius a having the origin as the center, and f is a function of $L^2(\Omega)$. The support of f is assumed to be contained in the interior of D_a.

Let Λ be a first order pseudo-diffrential operator defined on the boundary Γ_a of D_a which transforms the boundary data of the function u on Γ_a to the exterior normal derivative $\frac{\partial u}{\partial n}$ with respect to the exterior domain $\Omega_e = (D_a \cup \Gamma_a)^C$, where the function u is harmonic in Ω_e, being bounded at ∞. The operator Λ is said to be the exterior Steklov operator in this paper. Let $\Omega_i = \Omega \cap D_a$. Then the following interior problem $(\mathcal{E})$ is equivalent to (E):

$$(\mathcal{E})\quad \begin{cases} -\Delta u = f & \text{in } \Omega_i, \\ u = 0 & \text{on } \Gamma, \\ \dfrac{\partial u}{\partial n} = -\Lambda u & \text{on } \Gamma_a, \end{cases}$$

where the normal derivative $\frac{\partial}{\partial n}$ is taken with respect to the interior domain Ω_i.

Via the weak formulation (II) of ($\mathcal{E}$), the finite elelment approximation method can be applied to solve the problem, at least, formally, and abstractly. Namely the problem ($\mathcal{E}$) is equivalent to the following minimization problem (P):

$$(P) \qquad \min_{v\in H^1(\Omega_i),\ v|_\Gamma=0} \left[\frac{1}{2}\left(\int_{\Omega_i} |\nabla v|^2 dx + \int_{\Gamma_a} \Lambda u \cdot u d\Gamma\right) - \int_{\Omega_i} f v dx\right].$$

The quadratic form $\int_{\Gamma_a} \Lambda u \cdot u d\Gamma$ has an analytical expression through the Fourier series expansion. If we assume the existence of a family of finite element spaces approximating

$$V = \left\{v \in H^1(\Omega_i) : v|_\Gamma = 0\right\}$$

nicely, the constructed approximation problems may admit reasonable error estimates. The effect of the truncation of the Fourier series expansion of the operator Λ can be also discussed. A typical result is stated in Theorems 1 and 2 in this paper. They tell us that we may truncate the series with the number of terms, which is proportional to the inverse of the representative mesh length h of the finite element triangulation, in the viewpoint of the asymptotic convergence rate as h tends to 0.

Since the boundary of Ω_i is not polygonal, the approximation problem of the boundary is necessarily to be considered. One of the natural ways to approximate the circle Γ_a is to do it with the inscribed polygon with the equal length sides. As for this directions, Theorem 3 in §3 provides a justification for the case of piecewise linear continuous finite element spaces. The theory of curved elements due to Zlamal[12] and [13] plaies an essential role in the present investigation.

Finally the effect of the lumping procedure applied to the boundary bilinear form corresponding to the non-local boundary condition is considered. Theorem 4 in §4 tells us that there is no loss in the rate of convergence for the solutions considered in the Theorem in comparison with those considered in Theorems 1 and 3 if the solution of u belongs to $H^2(\Omega_i)$.

One of another numerical methods in the coupling of boundary element and finite element, which was well treated by the work Johnson-Nedelec[5]. A survey work [4] by Hsiao contains the development along this direction up to towards 1990. The technique of the present paper is naturally extended to more general elliptic problems in exterior domains, for example, the reduced wave equation or Helmholtz equation, which was treated by Masmoudi[8], in which, however, the clear information concerning the boundary approximation can not be found. More generally how to impose an appropriate boundary condition on the introduced artificial boundary is one of the interesting, and important subject in the numerical analysis of the problems in exterior domains. In this direction the authors owe to a survey work [3] by Hariharan, which contains the information up to towards 1985. The first author owes also to a domestic report [11] by Seto, which provides concrete analytical expressions of the exterior Steklov operators for various potential problems in exterior domains. The report gave the author his first step to investigate the problem concerned.

As for the construction of absorbing boundary conditions for the non-local boundary condition of the type like the condition on the artificial boundary considered in the paper, Feng originated a method called the natural boundary element method early (see [14], for example). He and his school have made a great progress in th application of this method. Some of the examples among such works are [14],[15] and [16]. The authors of the present paper admit that the philosophy of the work coincides with that of Feng and others in wide area, so that the conclusion of the paper may be considered very natural. They feel, however, that the peculiarity of this work is, at least, in the consideration of the effect of numerical integration for the boundary bilinear form, which is expressed through the truncated Fourier series expansions of the functions defined on the disc which is used as the artificial boundary in this work.

The length of the original manuscript with the proof is as about double as that of this paper. The authors hopes to publish the original manuscript near future.

§1. Setting of the problem and the result

To investigate (E), firstly we introduce the (exterior) Steklov operator. Let X be the real Hilbert space $L^2(\Gamma_a)$. Denote the inner product of X by (φ, ψ) for $\varphi, \psi \in X$. We have

$$(\varphi,\psi) = \int_{\Gamma_a} \varphi\psi \, d\Gamma \;=\; a\int_{-\pi}^{\pi} \varphi(a\cos\theta, a\sin\theta)\;\psi(a\cos\theta,\;a\sin\theta)\;d\theta.$$

The following system of functions is a complete orthonomal system of X.

$$\left\{\; C_0 = \frac{1}{\sqrt{2\pi a}},\; C_n = \frac{1}{\sqrt{\pi a}}\;\cos n\theta,\; S_n = \frac{1}{\sqrt{\pi a}}\;\sin n\theta,\; n = 1,2,\cdots \right\}.$$

Define the exterior Steklov operator Λ acting in X with the domain:

$$D(\Lambda) = \left\{ \varphi \in L^2(\Gamma_a) : \sum_{n=1}^{\infty} \left(\frac{n}{a}\right)^2 \left\{\; |(\varphi, C_n)|^2 + |(\varphi, S_n)|^2 \;\right\} < \infty \right\},$$

through the formula :

$$\Lambda\varphi = \sum_{n=1}^{\infty} \left(\frac{n}{a}\right) \{(\varphi, C_n)\, C_n + (\varphi, S_n)\, S_n\} \;,\quad \varphi \in D(\Lambda).$$

Consider the following exterior boundary value problem of the first kind for Laplace equation (E_e).

$$(E_e)\quad \left\{ \begin{array}{rcll} -\Delta u & = & 0 \;\text{in} & \Omega_e, \\ u & = & \varphi \;\text{on} & \Gamma_a, \\ \sup\limits_{\Omega_e} |u| & & < \infty. & \end{array} \right.$$

For any continous periodic function φ on Γ_a, there exists the unique classical solution u of (E_e) (see for example §35 of [10]). More precisely the solution u can be represented through

$$(1.1)\quad u = (\varphi, C_0)C_0 + \sum_{n=1}^{\infty} \left(\frac{a}{r}\right)^n \{(\varphi, C_n)\, C_n + (\varphi, S_n)\, S_n\} \quad \text{for } r > a.$$

Denote the exterior normal diffrentiation at the boundary Γ_a with respect to Ω_e by $\dfrac{\partial}{\partial n_e}$. If the data φ of the problem (E_e) belongs to $D(\Lambda)$, then for the solution u of (E_e) we have

$$(1.2)\quad \frac{\partial u}{\partial n_e} = \Lambda\varphi,$$

which is a consequence of the expression (1.1).

Introduce the following boundary value problem $(\mathcal{E})$.

$$(\mathcal{E})\quad \begin{cases} -\Delta u &= f \quad \text{in } \Omega_i, \\ u &= 0 \quad \text{on } \Gamma, \\ \dfrac{\partial u}{\partial n} &= -\Lambda u \quad \text{on } \Gamma_a. \end{cases}$$

Due to (1.1) and (1.2), we understand the equivalence of the problem (E) to $(\mathcal{E})$, at least formally. It is to be noted that a non-local boundary condition is imposed on the artificial boundary Γ_a.

To give the precise meaning of $(\mathcal{E})$, we understand it in the weak formulation as usual. To do so, we prepare the following notations:

$$\begin{aligned} a\,(u,v) &= \int_{\Omega_i} \nabla u\, \nabla v\; dx \quad \text{for } u,v \in H^1(\Omega_i), \\ F(v) &= \int_{\Omega_i} f\; v\; dx, \text{ for } f,v \in L^2(\Omega_i), \\ l\,(\varphi,\psi) &= (\Lambda^{1/2}\varphi,\ \Lambda^{1/2}\psi) \text{ for } \varphi,\psi \in H^{1/2}(\Gamma_a). \end{aligned}$$

It is well known that $H^{1/2}(\Gamma_a)$ coincides with $D(\Lambda^{1/2})$, where

$$D(\Lambda^{1/2}) = \left\{\varphi \in X : \sum_{n=1}^{\infty} \frac{n}{a}\left\{\, |(\varphi, C_n)|^2 + |(\varphi, S_n)|^2 \,\right\} < \infty\right\},$$

$$\Lambda^{1/2}\varphi = \sum_{n=1}^{\infty}\left(\frac{n}{a}\right)^{1/2} \{(\varphi, C_n)\, C_n + (\varphi, S_n)\, S_n\}\ .$$

Let γ be the trace operator from $H^1(\Omega_i)$ onto $H^{1/2}(\Gamma_a)$. Then we have the following problem (Π) as the weak formulation of $(\mathcal{E})$.

$$(\Pi)\quad \begin{cases} a(u,v) + l(\gamma u, \gamma v) = F(v), \quad v \in V, \\ u \in V. \end{cases}$$

Now we introduce an intermeditate problem (Π^M). Introduce bilinear forms $t_M(u,v)$ and $l_M(\varphi,\psi)$ as follows.

$$t_M(u,v) = a(u,v) + l_M(\gamma u,\ \gamma v), \quad l_M(\varphi,\psi) = l(P_M\varphi,\ P_M\psi),$$

where

$$P_M\varphi = (\varphi, C_0)\, C_0 + \sum_{n=1}^{M-1}\{\, (\varphi, C_n)\, C_n + (\varphi, S_n)\, S_n \,\} + (\varphi, C_M)\, C_M, \text{ for } \varphi \in X.$$

Then the problem (Π^M) is described below:

$$(\Pi^M)\quad \begin{cases} t_M(u,v) = F(v), \quad v \in V, \\ u \in V. \end{cases}$$

Let V_h be a finite dimensional subspace of V. Introduce the following discrete problem (Π_h^M), which is the standard Galerkin approximation of (Π^M).

$$(\Pi_h^M)\quad \begin{cases} t_M(u_h, v_h) = F(v_h), & v_h \in V_h, \\ u_h \in V_h. \end{cases}$$

Asuume there is a family $\{V_h : 0 < h \le \bar{h}\}$ of finite dimensional subspace V_h of V satisfying the following condition (h-1).

Condition (h-1) There is a constant C independent of $h \in (0, \bar{h}]$ and $v \in V \cap H^2(\Omega_i)$ such that

$$\min_{v_h \in V_h} \|v - v_h\|_{H^1(\Omega_i)} \le Ch\, \|v\|_{H^2(\Omega_i)}. \qquad \square$$

The problem (Π_h^M) can be considered as a resonable approximation problem of (Π) due to the following result.

Theorem 1 *Let u, and u_h^M, be the solution of (Π), and (Π_h^M), respectively. Suppose $u \in H^2(\Omega_i)$. Then there is a constant C independent of $h \in (0, \bar{h}]$ and u such that*

$$\|u - u_h^M\|_{H^1(\Omega_i)} \le C(h + \frac{1}{M})\|u\|_{H^2(\Omega_i)}. \qquad \square$$

§2. Error estimate for the solution of (Π_h^M)

Firstly we introduce the fractional power Λ^s of the Steklov operator Λ for any real s through

$$D(\Lambda^s) = \left\{\varphi \in L^2(\Gamma_a) : \sum_{n=0}^{\infty} \lambda_n^{2s}\left\{|(\varphi, C_n)|^2 + |(\varphi, S_n)|^2\right\} < \infty\right\},$$

$$\Lambda^s \varphi = \sum_{n=1}^{\infty} \lambda_n^{s} \{(\varphi, C_n)C_n + (\varphi, S_n)S_n\} \quad \text{for } \varphi \in D(\Lambda^s)$$

with

$$\lambda_n = \frac{n}{a}, \quad n = 0, 1, 2, \cdots.$$

Clearly we have $\Lambda^1 = \Lambda$. Denote the norm of $L^2(\Gamma_a)$ by $|\cdot|$:

$$|\varphi| = (\varphi, \varphi)^{1/2} = (\int_{\Gamma_a} |\varphi|^2 \, d\Gamma)^{1/2}.$$

The definition domain $D(\Lambda^s)$ of the operator Λ^s is a Hilbert space with the norm:

$$|\varphi|_s = (|\varphi|^2 + |\Lambda^s \varphi|^2)^{1/2} \quad \text{for } \varphi \in D(\Lambda^s),$$

which is denoted by $H^s(\Gamma_a)$. Let $H^m(D_a)$ be the usual Sobolev space with order m. According to the theory of interpolation spaces due to Lions-Magenes[7], we see that

$$H^{1+s}(D_a) = [H^2(D_a), H^1(D_a)]_{1-s}.$$

(If one follows [7] literally rigorously, we should consider that the element of $H^m(D_a)$ is complex valued. To avoid unnecessary explanations concerning this point, we adopt a conventional introduction of the interpolation theory. We use freely the result of [7].).

Auxiliarly we introduce the following interpolation space V^{1+s} for $s \in [0,1]$:

$$V^{1+s} = [H^2(\Omega_i) \cap V, V]_{1-s},$$

where $H^2(\Omega_i) \cap V$, and V, are the Hilbert spaces with the norms of $H^2(\Omega_i)$, and $H^1(\Omega_i)$, respectively. Since the identity mappings form $H^2(\Omega_i) \cap V$ into $H^2(\Omega_i)$, and from V into $H^1(\Omega_i)$, are continuous, we have the fact that V^{1+s} is continuously imbeded into $H^{1+s}(\Omega_i)$.

Now we have the following result, which implies Theorem 1 as a special case.

Theorem 2 *Let u be the solution of* (II). *Suppose $u \in V^{1+s}$ for some $s \in [0,1]$. Assume condition (h-1). Then there is a constant C independent of $s \in [0,1]$ and $u \in V^{1+s}$ such that*

$$\|u - u_h^M\|_{H^1(\Omega_i)} \leq C(h^s + M^{-s}) \, \|u\|_{V^{1+s}}. \qquad \square$$

At present the authors do not know whether V^{1+s} can be replaced with $H^{1+s}(\Omega_i)$ in the statement of Theorem 2 or not, and a characterization of V^{1+s} which is useful for practical purpose, especially in the case of planar domain $\mathcal{O}$ with corners.

§3. Approximation of the boundary

Now we approximate the boundary curves Γ, and Γ_a, with polygonal curves Γ^M, and Γ_a^M, respectively. If Γ is piecewise smooth, then all the non-smooth points of Γ should be located at the corners of the polygonal curve Γ^M. The polygonal curve Γ_a^M is the boundary of the regular polygon with $2M$ equally lengthy sides, and it is inscribed to the artificial boundary Γ_a. (See Figure 1 in the case of M = 6.) Let the domain between Γ_a^M and Γ^M be Ω_i^M, which is triangulated into a set $\mathcal{T}^M$ of the element triangles T in such a way that all the vertices of the curves Γ_a^M and Γ^M are the vertices of element triangles, and that there are no other vertices of element triangles exist on the curve Γ_a^M.

Let $a^M(u,v)$ be the Dirichlet inner product on the domain Ω_i^M:

$$a^M(u,v) = \int_{\Omega_i^M} (\nabla u, \nabla v) \mathrm{dx}.$$

Let $b_j, j = 0, \pm 1, \cdots, \pm(M-1), M$ be the nodal points on the artificial boundary with respect to the above defined triangulation $\mathcal{T}^M$. For simplicity of presentation, assume that b_j is represented as

$$(3.1) \quad \begin{cases} b_j &= (a\cos\theta_j, a\sin\theta_j), \\ \theta_j &= \dfrac{j\pi}{M}, \\ & (j = 0, \pm 1, \cdots, \pm(M-1), M). \end{cases}$$

Let $\widehat{W}^M$ be the piecewise linear continuous finite element spcace associated with the triangulation $\mathcal{T}^M$:

$$\widehat{W}^M = \left\{ \hat{v}_h : \hat{v}_h \in C(\overline{\Omega}_i^M), \ \hat{v}_h|_T \in \mathbf{P}_1 \ \text{ for } \forall T \in \mathcal{T}^M \right\},$$

where $\mathbf{P}_1$ is the totality of polynomials with degree at most 1. Define $\hat{V}^M$ through

$$\hat{V}^M = \left\{\hat{v}_h \in \widehat{W}^M : \hat{v}_h|_{\Gamma^M} = 0\right\}.$$

Now we introduce the following bilinear form $\hat{l}^M(\hat{u}_h, \hat{v}_h)$ defined on $\widehat{W}^M$ through

$$\hat{l}^M(\hat{u}_h, \hat{v}_h) = \sum_{i,j=-(M-1)}^{M} l_{ij}\hat{u}_h(b_j)\hat{v}_h(b_i)$$

where

$$(3.2) \quad \begin{cases} l_{ij} = \dfrac{\pi a}{M^2}\left\{\sum_{n=1}^{M-1} \lambda_n \alpha(\frac{n\pi}{M})^2 \cos\left((i-j)\frac{n}{M}\pi\right) + (-1)^{i+j}\lambda_M \alpha(\pi)^2\right\} \\ \text{with} \\ \lambda_n = \dfrac{n}{a} \text{ and } \alpha(\theta) = \dfrac{2(1-\cos\theta)}{\theta^2}. \end{cases}$$

Let f^M be an element of $L^2(\Omega_i^M)$. Define

$$\hat{F}^M(\hat{v}_h) = \int_{\Omega_i^M} f^M \hat{v}_h \, dx.$$

Now we can state our computable discrete problem $(\hat{\Pi}^M)$ as the following:

$$(\hat{\Pi}^M) \quad \begin{cases} a^M(\hat{u}_h, \hat{v}_h) + \hat{l}^M(\hat{u}_h, \hat{v}_h) = \hat{F}^M(\hat{v}_h), & \hat{v}_h \in \hat{V}^M, \\ \hat{u}_h \in \hat{V}^M. \end{cases}$$

Since the bilinear form $a^M(\hat{u}_h, \hat{v}_h) + \hat{l}^M(\hat{u}_h, \hat{v}_h)$ is coercive on $\hat{V}^M$, the problem $(\hat{\Pi}^M)$ has the unique solution.

For simplicity of the presentation we assume that the interior boundary Γ is either (a) a union of the finite number of C^3 curves or (b) a polygonal curve. Assume further that f has the following property (F):

$$(F) \quad \begin{cases} \text{If } \Gamma \text{ is of the case (a), there is an open set } \mathcal{O}_0 \text{ with the closure } \overline{\mathcal{O}_0} \\ \text{which is contained in } \mathcal{O} \text{ such that} \\ f \in H^2(\Omega_0) \text{ with } \Omega_0 = {\mathcal{O}_0}^C \cap D_a. \\ \text{If } \Gamma \text{ is of the case (b),} \\ f \in H^2(\Omega_0) \text{ with } \Omega_0 = \Omega. \end{cases}$$

Let $b_j, j = 1, 2, \cdots, N$, be the nodal points in the closure Ω_i^M associated to the triangulation $\mathcal{T}^M$, and let $\hat{w}_j$ be the basis function of $\widehat{W}^M$ with the property that

$$\hat{w}_i(b_j) = \delta_{ij} \quad 1 \le i, j \le N.$$

Then we adopt the interpolation function of f as f^M. Namely we use

$$f^M = \sum_{j=1}^{N} f(b_j)\hat{w}_j.$$

Now we specify a method of constructing the approximate interior boundary Γ^M. Fix a positive integer $\underline{M}$. For any $M \geq \underline{M}$, let $L = L(M)$ be a positive integer, and let $h = h(M)$ be a positive number such that

$$C_1 M \leq L(M) \leq C_2 M,$$

and such that

$$C_3 M^{-1} \leq h(M) \leq C_4 M^{-1},$$

respectively, where C_i $(i = 1,2,3,4)$ are positive constants independent of M. Choose L points on the boundary Γ such a way that each of the end points of the C^3-portions of Γ is one of the L points, and that the maximum of the distances of two points of them, being located consecutively along the curve Γ, is less than or equal to h. Let Γ^M be the closed polygonal line joining all the L points consecutively along the curve Γ.

Now we triangulate the domain Ω_i^M such that the totalities of the nodal points on Γ_a^M, and on Γ^M, coincide with the $2M$ end points, and the L end points chosen as above, respectively, and that the maximum of the diameters of the element triangles is less than or equal to $C_5 h$ where C_5 is a positive constant independent of M. Further we require that there are no element triangles all of whose vretices are boundary nodal points. Denote the triangulation mentioned as above by $\mathcal{T}^M$. We understand that $h = h(M)$ is the representive mesh length of $\mathcal{T}^M$. Assume further the family $\{\mathcal{T}^M : M \geq \underline{M}\}$ is regular and satisfies an inverse assumption in the sense defined in p.147 of [2]. Namely there exist two constants σ and ν such that

$$\frac{h_T}{\rho_T} \leq \sigma, \ \frac{h}{h_T} \leq \nu \ \text{ for any } T \in \bigcup_{M \geq \underline{M}} \mathcal{T}^M,$$

where h_T, and ρ_T, are the diameter of T, and the diameter of the inscribed circle of T respectively, and $h = h(M)$.

Under the above setting, we have a sequence of approximate problems $(\widehat{\Pi}^M)$.

Now we correspond $\widehat{W}^M$, and $\widehat{V}^M$, to W^M and V^M according to the curved element techniquedue due to Zlamal [12] and [13], where W^M, and V^M, are subspaces of $H^1(\Omega)$, and V, respectively.

The following is an introduction of Zlamal's curved elements with a suitable modification for our circular boundary. Let T be an element of $\mathcal{T}^M$ one of whose sides is a part of boundary $\Gamma^M \cup \Gamma_a^M$. The totality of such element triangles is denoted by $\partial\mathcal{T}^M$. Let the vertices of T be $b_j, j = 0,1,2$, such that the segment $\overline{b_0 b_2}$ is a part of the boundary of Ω. Let $\hat{T}$ be the reference triangle with the vertices $\hat{b}_j$ $(j = 0,1,2)$:

$$\hat{b}_0 = (0,0)^T, \ \hat{b}_1 = (0,1)^T, \ \hat{b}_2 = (1,0)^T.$$

For the notational convenience, the points $b, x, \hat{b}, \xi, \cdots$ in R^2 are identified with the 2-dimensional column vectors $b = (b^1, b^2)^T$, $x = (x^1, x^2)^T$, $\hat{b} = (\hat{b}_1, \hat{b}_2)^T$, $\xi = (\xi^1, \xi^2)^T, \cdots$. Then the mapping:

$$x_0(\xi) = b_0 + \xi^1(b_1 - b_0) + \xi^2(b_2 - b_0)$$

maps bijectively $\hat{T}$ onto T. Assume the portion of the boundary curve between b_0 and b_2 is represented as

$$\varphi(s) = (\varphi^1(s), \varphi^2(s))^T, \quad s_0 \le s \le s_2,$$

where $\varphi(s_0) = b_0$ and $\varphi(s_2) = b_2$. If b_0 and b_2 are the points on Γ_a;

$$b_0 = a(\cos\theta_0, \sin\theta_0)^T, \quad b_2 = a(\cos\theta_2, \sin\theta_2)^T$$

with $\theta_0 > \theta_2$, choose φ as

$$\varphi(s) = a(\cos(\theta_0 - s), \sin(\theta_0 - s))^T, \quad s_0 \le s \le s_2$$

with $s_0 = 0$ and $s_2 = \theta_0 - \theta_2$. For $\xi^2 \in \mathrm{R}$, set

$$s(\xi^2) = (1 - \xi^2)s_0 + \xi^2 s_2, \quad b(\xi^2) = (1 - \xi^2)b_0 + \xi^2 b_2,$$

and

$$\Phi(\xi^2) = \frac{1}{1 - \xi^2}\left(\varphi(s(\xi^2)) - b(\xi^2)\right).$$

Finally define the mapping:

$$x(\xi) = x_0(\xi) + (1 - \xi^1 - \xi^2)\Phi(\xi^2).$$

For $\underline{M}$ large enough, the mapping $x(\xi)$ maps bijectively $\hat{T}$ onto $\widetilde{T}$ for any $T \in \partial\mathcal{T}^M$ with $M \ge \underline{M}$, where $\widetilde{T}$ is the region surrounded by the segments $\overline{b_0 b_1}$ and $\overline{b_1 b_2}$, and the portion of the boundary curve between b_2 and b_0. In the sequel $\widetilde{T}$ is said to be a curved element.

For any $\hat{v}_h \in \widehat{W}^M$, define a function $\widetilde{v}_h$ on Ω_i through the following way:

$$(3.3) \quad \begin{cases} \text{if } T \in \mathcal{T} - \partial\mathcal{T}^M, & \widetilde{v}_h\,|_T = \hat{v}_h, \\ \text{if } T \in \partial\mathcal{T}^M, & \widetilde{v}_h\,(x(\xi)) = \hat{v}_h(x_0(\xi)) \quad \text{for } \xi \in \hat{T}. \end{cases}$$

The totality of such functions $\widetilde{v}_h$ is denoted by W_h, and the totality of $\widetilde{v}_h$ with $\hat{v}_h \in \hat{V}^M$ is denoted by V_h, with the correspondence of $h = h(M)$. The space W_h, and V_h, are the subspace of $H^1(\Omega_i)$, and V, respectively. We have the following result.

Theorem 3 *Suppose f has the property (F). Let u and u_h be the solution of (Π), and $(\hat{\Pi}^M)$, respectively. If $u \in H^2(\Omega_i)$, then there is a constant C independent of $M \ge \underline{M}$ and u such that*

$$\|u - \widetilde{u_h}\|_{H^1(\Omega_i)} \le Ch\left(\|u\|_{H^2(\Omega_i)} + \|f\|_{H^2(\Omega_0)}\right)$$

where $\widetilde{u_h}$ is determined from $\hat{u}_h$ through (3.3). □

§4. Effect of the lumping procedure for the non-local boundary condition

Firstly we introduce the following bilinear form with the argument φ and ψ being functions defined on Γ_a.

$$\bar{l}_M(\varphi,\psi) = \sum_{i,j=-(M-1)}^{M} \bar{l}_{ij}\varphi(b_j)\psi(b_j),$$

where

$$\bar{l}_{ij} = \frac{\pi a}{M^2}\left\{\sum_{n=1}^{M-1}\lambda_n \cos\left((i-j)\frac{n}{M}\pi\right) + (-1)^{i+j}\lambda_M\right\} \text{ with } \lambda_n = \frac{n}{a},$$

and b_j, $j = 0, \pm 1, \cdots, \pm(M-1), M$, are defined by (3.1). It holds

$$\bar{l}_{ij} = \begin{cases} \dfrac{\pi}{2}\dfrac{M-1}{M} & : \ i = j, \\ \dfrac{\pi}{2M^2}\left[(-1)^{i-j}M - \dfrac{1-(-1)^{i-j}}{1-\cos\frac{(i-j)\pi}{M}}\right] & : \ i \neq j \end{cases}$$

(see [1]). The introduction of the boundary bilinear form $\bar{l}_M(\gamma u_h, \gamma v_h)$ can be considered as a lumping procedure for the non-local boundary condition.

For $h = h(M)$, denote $a^M(u,v)$ by $a_h(u,v)$. Introduce the bilinear form $a_h^*(\widehat{u_h}, \widehat{v_h})$ on V_h through

$$a_h^*(\widehat{u_h}, \widehat{v_h}) = a_h(\hat{u}_h, \hat{v}_h) \quad \text{for } \widehat{u_h}, \widehat{v_h} \in V_h,$$

where the correspondence: $\widehat{v_h} \longleftrightarrow \hat{v}_h$ is given by (3.3). The following problem $(\overline{\Pi}_h^*)$ is considered in this section.

$$(\overline{\Pi}_h^*) \quad \begin{cases} a_h^*(u_h, v_h) + \bar{l}_M(\gamma u_h, \gamma v_h) = F(v_h), \quad v_h \in V_h, \\ u_h \in V_h, \end{cases}$$

where V_h is the same as that determined in §3.

Theorem 4 *Let u, and u_h, be the solution of* (Π), *and* $(\overline{\Pi}_h^*)$. *If $u \in H^2(\Omega_i)$, then there is a constant C independent of $M \geq \underline{M}$ and u such that*

$$\|u - u_h\| \leq Ch\left(\|u\|_{H^2(\Omega_i)} + |u''|_{L^2(\Gamma_0)}\right),$$

where

$$u'' = \frac{\partial^2}{\partial\theta^2}u(a\cos\theta, a\sin\theta). \qquad \square$$

Since the support of f is contained in D_a, Weyl-Schwartz lemma implies that u is smooth in a neighbourhood of Γ_a. Hence the quantity $|u''|_{L^2(\Gamma_0)}$ is meaningful, and finite for the solution u of (Π). The discussion on the effect of the numerical integration for F is omitted here for simplicity of the presentation.

§5. Report of the numerical experience

5.1 A numerical test. In the computation we employed a square with its side length being 2 as the obstacle $\mathcal{O}$. More precisely

$$\mathcal{O} = \{(x_1, x_2) : -1 < x_1, x_2 < 1\}.$$

Take f as

$$f = \begin{cases} 1, & : r \leq 3, \\ 0, & : r > 3, \end{cases}$$

where $r = (x_1{}^2 + x_2{}^2)^{1/2}$. Let $a = 5$. Set $\underline{M} = 12$. The region $\Omega_i^{\underline{M}}$ is triangulated as in the Figure 2. For $M = 2^p\underline{M}$ with $p = 1, 2, 3$, a refinement of $\mathcal{T}^{2^{-1}M}$ is employed for the triangulation $\mathcal{T}^M$. More precisely, every triangular element T of $\mathcal{T}^{2^{-1}M}$ is divided into four congruent small triangles using the vertices of T and three mid points of the sides of T except for the case of T having two vertices on the artificial boundary Γ_a, in which case, instead of T, the quadrilateral, whose sides join the three vertices of T and the new nodal point lying on Γ_a between two vertices of T, is divided into 4 small triangles (see Figure 3).

For $M = 3\underline{M}$, we construct the triangulations $\mathcal{T}^M$ as a refinement of $\mathcal{T}^{\underline{M}}$, in which case every $T \in \mathcal{T}^{\underline{M}}$ is divided into nine congruent small triangles using the vertices of T, and six one-third equi-divided points of the sides of T, and the center of the gravity of T except for the case of T having two vertices on the artificial boundary Γ_a, in which case, instead of T, the pentagon whose sides join the three vertices of T and the two new nodal points lying on Γ_a between two vertices of T, is divided into 9 small triangles. Then $\mathcal{T}^{6\underline{M}}$ is constructed from $\mathcal{T}^{3\underline{M}}$ through the procedure from $\mathcal{T}^{\underline{M}}$ to $\mathcal{T}^{2\underline{M}}$ as mentioned before. By this method, we have constructed a set of triangulation $\{\mathcal{T}^{p\underline{M}} : \underline{M} = 12, p = 1, 2, 3, 4, 6, 8\}$. Let $\widehat{D}_3$ be the union of all triangles $T \in \mathcal{T}^{12}$ contained in D_3. Let

$$f^M = \begin{cases} 1, & : (x, y) \in \widehat{D}_3, \\ 0, & : otherwise, \end{cases}$$

and let

$$\widehat{F}^M(\widehat{v}_h) = \int_{\Omega_i{}^M} f^M \widehat{v}_h dx.$$

Consider the problem $(\Pi_h{}^*)$ replacing $F(v_h)$ with $F^M(v_h)$. Then we see that the problem $(\Pi_h{}^*)$ is the same problem as the problem $(\overline{\Pi}^M)$ stated below:

$$(\overline{\Pi}^M) \quad \begin{cases} a^M(\widehat{u}_h, \widehat{v}_h) + \overline{l}^M(\widehat{u}_h, \widehat{v}_h) &= \widehat{F}^M(\widehat{v}_h), \quad \widehat{v}_h \in \widehat{V}_h, \\ \widehat{u}_h \in \widehat{V}_h, \end{cases}$$

where

$$\overline{l}^M(\widehat{u}_h, \widehat{v}_h) = \sum_{i,j=-(M-1)}^{M} \overline{l_{ij}} \widehat{u}_h(b_i) \widehat{v}_h(b_j).$$

Our computational work was done for the problem $(\overline{\Pi}^M)$ settled as above. Let $u^M(x, y)$ be the solution of $(\overline{\Pi}^M)$. Our numerical solution were obtained with the Sun SPARC Station 2 installed in Department of Computer Science and Information Mathematics, The University of Electro-Communications. We have used FORTRAN 77 language in double precision arithmetic. Figures 4, and 5, show the dependence of M for the profile $u^M(r,0)$, and that of $u^M(\frac{\sqrt{2}}{2}r, \frac{\sqrt{2}}{2}r)$, respectively.

Although these figures show a good convergence result, an asymptotic theoretical error estimate for the present numerical test problem has not yet established so far, since we have asuumed that the solution $u \in H^2(\Omega_i)$ in Theorem 4.

5.2 A practical method of the numerical evaluation of the rate of convergence. First of all we remark the following Theorem which is easy to be proven (Cf. [6]).

Theorem 5 *Let X be a Banach space with the norm $\|\cdot\|$. Fix an element u of X. Suppose for any $h \in (0, \overline{h}]$ with $\overline{h} < \infty$, there is an element $u(h)$ of X. Define numerical functions $e(h)$ and $\tilde{e}(h)$ through*

$$e(h) = \|u(h) - u\|, \ \tilde{e}(h) = \|u(h) - u(h/2)\|.$$

Assume there are positive constants $\alpha, c_1, c_2, \tilde{\alpha}, \tilde{c}_1, \tilde{c}_2$ such that

$$c_1 h^\alpha \le e(h) \le c_2 h^\alpha, \ \tilde{c}_1 h^{\tilde{\alpha}} \le \tilde{e}(h) \le \tilde{c}_2 h^{\tilde{\alpha}}$$

hold for $h \in (0, \overline{h}]$. Then we have $\alpha = \tilde{\alpha}$. □

Based on Theorem 5, we understand that it is very plausible that we conjecture the rate of convergence of $e(h)$ from the numerical values of $\tilde{e}(h)$ with several sufficiently small values of h. We have tried to calculate the values of the type:

$$\tilde{e}(M) = \|u^M - u^{2M}\|$$

for our obtained numerical solutions u^M. The norms employed are L^2 norm, Dirichlet norm and H^1 norm.

If we have the inclusion

$$(5.1) \quad \hat{V}^M \subset \hat{V}^{2M} \subset H^1(\Omega_i),$$

u^M is an element of $\hat{V}^{2M}$, and we can compute the value $\tilde{e}(M)$ using the stiffness and the mass matrices corresponding to the triangulation $\mathcal{T}^{2M}$. In our case, condition (5.1) does not hold. Define a standard counterpart $\tilde{u}^M$ in $\hat{V}^{2M}$ of u^M in $\hat{V}^M$ by the following way. Let the boundary nodal points on Γ_a^M be

$$b_j{}^M = a(\cos j\frac{\pi}{M}, \sin j\frac{\pi}{M}), \quad j = 0, \pm 1, \pm 2, \cdots, \pm(M-1), M.$$

(See Figure 3.) For all the nodal points b of $\mathcal{T}^{2M}$ except for the nodal points of the form $b_{2j+1}{}^{2M}$, define

$$\tilde{u}^M(b) = u^M(b),$$

and for the points $b_{2j+1}{}^{2M}$, define

$$\tilde{u}^M(b_{2j+1}{}^{2M}) = \frac{u^M(b_j{}^M) + u^M(b_{j+1}{}^M)}{2}.$$

Then an element $\tilde{u}^M \in V^{2M}$ is uniquely determined. For such $\tilde{u}^M$, we have computed the value of $\|\tilde{u}^M - u^{2M}\|$. Let

$$\begin{aligned}\tilde{e}_L(M) &= \|\tilde{u}^M - u^{2M}\|_{L^2(\Omega_i^{2M})},\\ \tilde{e}_H(M) &= \|\tilde{u}^M - u^{2M}\|_{H^1(\Omega_i^{2M})},\\ \tilde{e}_E(M) &= \|\nabla(\tilde{u}^M - u^{2M})\|_{L^2(\Omega_i^{2M})}.\end{aligned}$$

In Figure 6, the computed values of $\tilde{e}_L(M), \tilde{e}_H(M)$, and $\tilde{e}_E(M)$ are shown using the log-log scale. We employed for $M = 12, 24, 36, 48$. The lines in the Figure are the graphs of the linear functions $y = -\alpha x + \beta$ which are the least square approximations of the values $(x, y) = (\log M, \log \tilde{e}(M))$, $M = 12p$, $p = 1, 2, 3, 4$. We obtained

$$\alpha = \begin{cases} 1.364 & \text{for} \quad \tilde{e} = \tilde{e}_L, \\ 0.838 & \text{for} \quad \tilde{e} = \tilde{e}_H, \\ 0.625 & \text{for} \quad \tilde{e} = \tilde{e}_E. \end{cases}$$

And the gradients α of the lines which connect the two points corresponding $M = 36$ and $M = 48$ is the following:

$$\alpha = \begin{cases} 1.369 & \text{for} \quad \tilde{e} = \tilde{e}_L, \\ 0.750 & \text{for} \quad \tilde{e} = \tilde{e}_H, \\ 0.659 & \text{for} \quad \tilde{e} = \tilde{e}_E. \end{cases}$$

Our calculation suggests that the solution u^M converges to an element

$$u \in H^{1+s}(\Omega_i), \quad s = 0.66\cdots.$$

This conjecture seems to be compatible with the shape of the obstacle $\mathcal{O}$. Namely Ω has the four interior corners with angles $\frac{3}{2}\pi$.

Any way both the computation through the scheme $(\widehat{\Pi}^M)$, and the theoretical study for the problem $(\overline{\Pi}^M)$, with the assumption of $u \in V^{1+s}$ and/or $u \in H^{1+s}(\Omega_i)$, are left to our future subjects.

Acknowledgement

The first author should like to express his sincere gratitude to Professors Feng Kang, Yu De-Hao of Computing Center, Academia Sinica, who kindly gave him the valuable information on their works in the direction of the present work during the conference. The authors express their hearty thanks to Professor T. Kako of the University of Electro-Communications for his instructions of references, discussions with them, and encouragement to them. They thank also Mr. S. Nagoya of Ph.D student of the University of Electro-Communications who gave them an explicit form of $\bar{l}_{ij}$ mentioned in §4.

References

[1] Ajiro, A., *Numerical computation for the Poisson equation in an exterior domain*, Master thesis, in Japanese, Course of Computer Science and Information Mathematics, The University of Electro-Communications (March, 1992).

[2] Ciarlet, Ph., *The Finite Element Method for Elliptic Problems*, North-Holland, Amsterdam,NewYork,Oxford, 1978.
[3] Hariharan, S. I., Absorbing boundary conditions for exterior problems, Chapter 6 of Numerical Methods for Partial Differential Equations, Ed. by Hariharan, S. I., Moulden, T. H., Pitman research notes in mathematics; 145, Longman Scientific & Technical, Harlow, 1986.
[4] Hsiao, G. C., The coupling of boundary element and finite element methods, *Z. Angew. Math. Mech.* **70**, 493–503 (1990).
[5] Johnson, C., and Nedelec, J. C., On the coupling of boundary integral and finite element methods, *Math. Comp.*, **35**, 1063–1079 (1980).
[6] Kurata, S., Ushijima, T., Behaviors of errors of finite element numerical solutions of 3-D potential problems computed with iterative methods, (in Japanese), Advances in Numerical Methods for Large Sparse Sets of Linear Equations. Ed. by Natori, M. and Nodera, T., *Proceedings of the 6th Symposium on Large Sparse Sets of Linear Equations "Parallel Processing for Scientific Computing"*, March 1, 1990, Keio University.
[7] Lions, J. L., and Magenes, E., *Non-Homogeneous Boundary Value Problem and Applications I*, Springer, Berlin (1972).
[8] Masmoudi, M., Numerical solution for exterior problem, *Numer. Math.*, **51**, 87–101(1987).
[9] Nečas, J., *Les Méthodés Directes en Théorie des Équations Elliptiques*, Masson et Cie, Paris & Academia, Plague (1967).
[10] Petrovsky, I. G ., *Lecture on Partial Differential Equations*, Russian original, National Publisher of Technical Theoretical Literatures, Moscow, 1953, Japanese translation, Tokyo-Tosyo, Tokyo, 1958.
[11] Seto, H., Open boundary conditions in scalar potential fields and their application, (in Japanese), presented at the 7th Pan-Setouchi C&A Seminar, held in Ube, July 1987.
[12] Zlamal, M., Curved elements in the finite element method. I, *SIAM J. Numer. Anal.*, **10**, No.1, 229–240 (1973).
[13] Zlamal, M., Curved elements in the finite element method. II, *SIAM J. Numer. Anal.*, **11**, No.2, 347–362 (1974).

Supplementary References

[14] Feng, K., Finite element method and natural boundary reduction, *Proceedings of the International Conference of Mathematicians*, August 16-24, 1983, Warszawa, 1439–1453 (1983).
[15] Han, H., Wu, X., A finite element method for linear elastic equations on unbounded domains, *Proceedings of International Conference on Scientific Computing*, Hangzhou, China August 20-23, 1991, Ed. by Chan, T., Shi, Z., World Scientific, 36–40 (1992).
[16] Yu, D., Approximation of boundary conditions at infinity for harmonic equation, *Journal of Computational Mathematics*, **3**, No.3, 219–227 (1985), published by Science Press, Beijing.

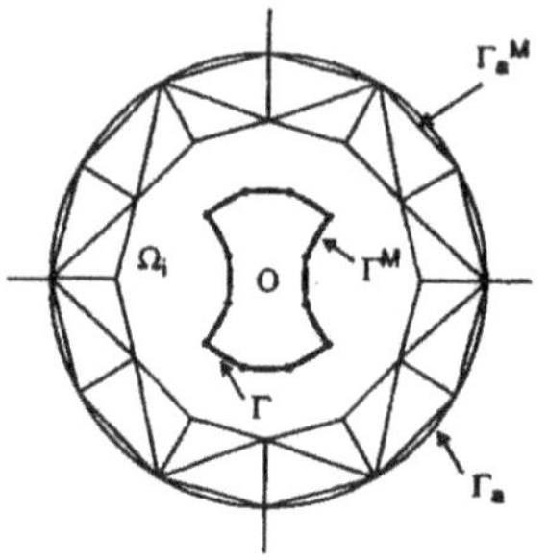

Figure 1 Boundary approximation.

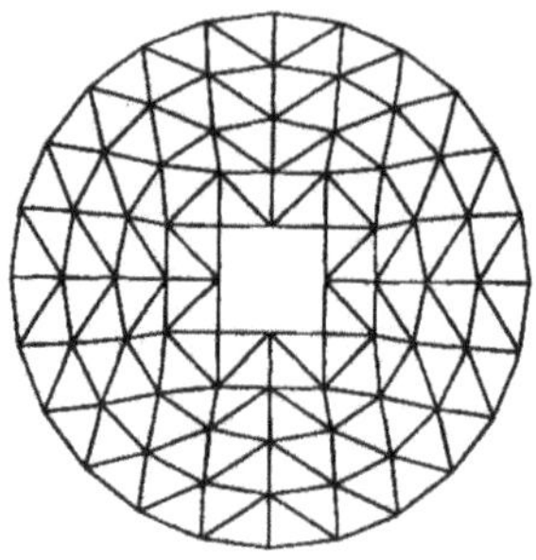

Figure 2 Basic triangulation of Ω_i^M with $M = 12$.

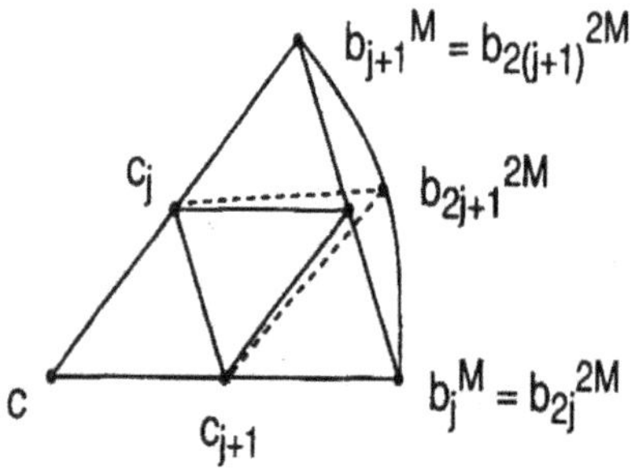

Figure 3 Refinement of boundary triangle.

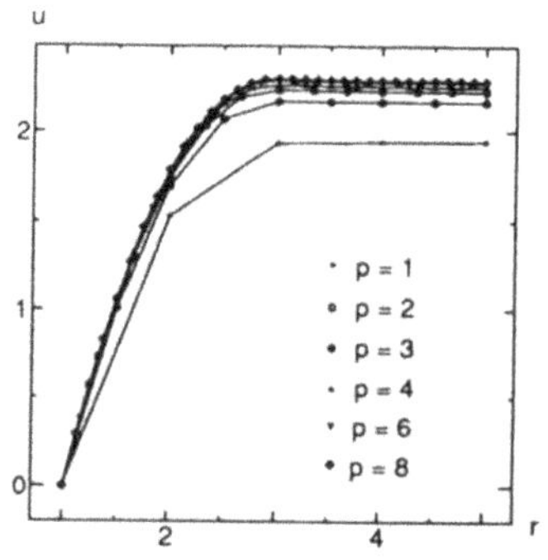

Figure 4 Profile of $u^M(r, 0)$.

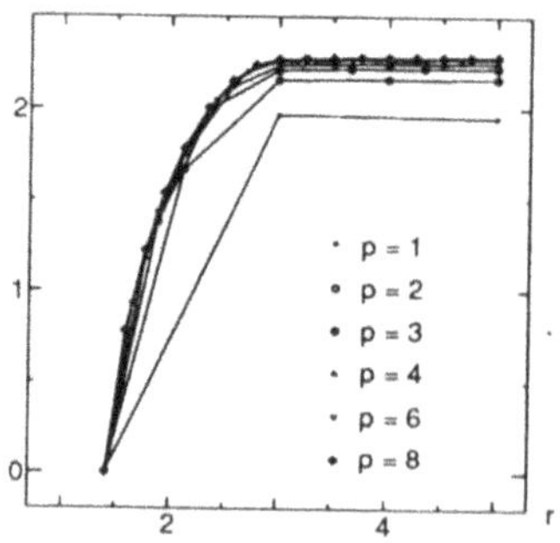

Figure 5 Profile of $u^M(\frac{\sqrt{2}}{2}r, \frac{\sqrt{2}}{2}r)$.

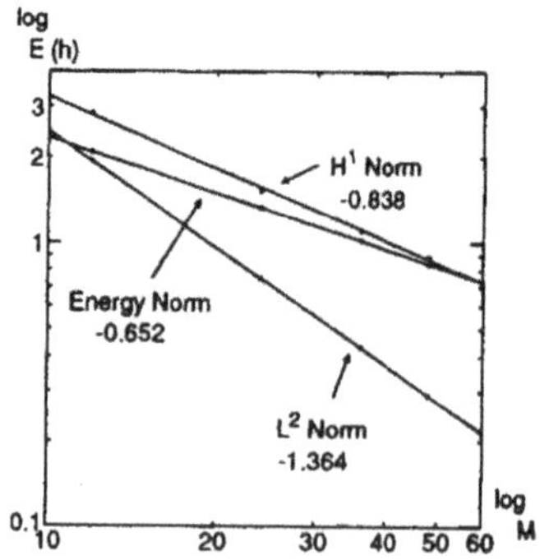

Figure 6 Rate of convergence.

A PANORAMIC VIEW ON KINK–IMPURITY INTERACTIONS IN THE SINE-GORDON AND ϕ^4 MODELS

ZHANG FEI, YURI S. KIVSHAR† and LUIS VAZQUEZ
Departamento de Física Teórica I, Facultad de Ciencias Físicas, Universidad Complutense, E-28040 Madrid, Spain

†*Institute für Theoretische Physik*
Universität Düsseldorf, D-4000, Düsseldorf, Germany

ABSTRACT

A panoramic view on recent results concerning interactions of kinks with point impurities is presented for the sine-Gordon and ϕ^4 models. The resonance structures arised in the scattering of kink by an attractive impurity are analyzed. A brief summary of the results about two and many-impurity effects is also presented.

1 Introduction

In recent years, wave propagation through inhomogeneous and disordered media has attracted increasing attention [1]-[3]. One of the central issues is how the disorder may change the steady-state motion of solitons. As a first step in this rich program, we have studied the kink dynamics in the presence of one, two and many point impurities [4]-[10].

As it is well known, there are, in general, three main types of solitons: dynamical solitons, envelope solitons, and topological solitons (kinks). The amplitude of a dynamical soliton is proportional to its velocity and, when propagating along a disordered system, their scattering is similar to that of a linear wave packet [11]. Envelope solitons are two-parametrical ones, which exhibit a quite complicated behavior [12]. The last type of solitary waves, kinks, are topologically stable and they resemble classical particles. Therefore, dynamics of kinks can be usually described by the so-called collective-coordinate approach.

In the present contribution we briefly review our recent results concerning the kink dynamics under influence of impurities. In Section 2 and 3 we analyze the kink interactions with one impurity in the sine-Gordon (SG) and ϕ^4 model respectively. In Section 4 we discuss two and many impurity effects on the kink dynamics. We draw conclusions and indicate some interesting open questions in Section 5.

2 Kink-impurity interactions in the SG model

Let us consider the SG equation including a local inhomogeneity,

$$u_{tt} - u_{xx} + [1 - \epsilon\delta(x)] \sin u = 0, \tag{1}$$

where $\delta(x)$ is the Dirac δ-function that can be considered as the limit of the following continuously differentiable function,

$$\delta(x, \alpha) = \frac{\alpha}{\sqrt{\pi}} \exp(-\alpha^2 x^2). \tag{2}$$

When the perturbation is absent ($\epsilon = 0$), the SG model (1) has kink solutions, the so-called topological solitons,

$$u_k(x, t) = u_k(z) = 4 \tan^{-1} \exp[(x - X)/\sqrt{1 - V^2}], \tag{3}$$

where $X = Vt + X_0$ is the kink coordinate, $\dot{X} = V$ being its velocity.

If $\epsilon > 0$ Eq. (1) supports a localized impurity mode:

$$u_{im}(x, t) = a_0 \cos(\Omega t + \theta_0) e^{-\epsilon|x|/2}, \tag{4}$$

where Ω represents the frequency of the impurity mode,

$$\Omega = \sqrt{1 - \epsilon^2/4} \tag{5}$$

and θ_0 is an arbitrary initial phase.

We discretize Eq.(1) with a conservative numerical scheme and carry out the simulations in the spatial interval $(-60, 60)$ with discrete stepsizes $\Delta x = 2\Delta t = 0.04$. When handling the Dirac $\delta-$function, we take its value equal to $1/\Delta x$ at $x = 0$, and zero otherwise. The initial conditions are always taken as a kink centered at $X = -6$, moving toward the impurity with a given initial velocity V_i. We have made intensive numerical simulations of the problem for $\epsilon > 0$ (attractive impurity), and the main results may be summarized as follows:

- There exists a critical velocity $V_c(\epsilon)$, such that if $V_i > V_c$, the kink will pass the impurity inelastically and escape to the positive infinity, losing part of its kinetic energy through radiation and excitation of an impurity mode. The final kink velocity V_f can be estimated as $V_f^2 = \alpha(\epsilon)(V_i^2 - V_c^2)$, where $\alpha(\epsilon)$ is determined from numerical data.

- If $V_i < V_c$, the kink cannot escape to infinity after the first interaction, but it will stop, and then come to interact with the impurity again. For most velocities, the kink will lose energy again in the second interaction and finally get trapped by the impurity. However, for some special incoming velocities, the kink may escape to the negative infinity after the second interaction, i.e., the kink may be totally *reflected* by the attractive impurity (see Fig. 1)

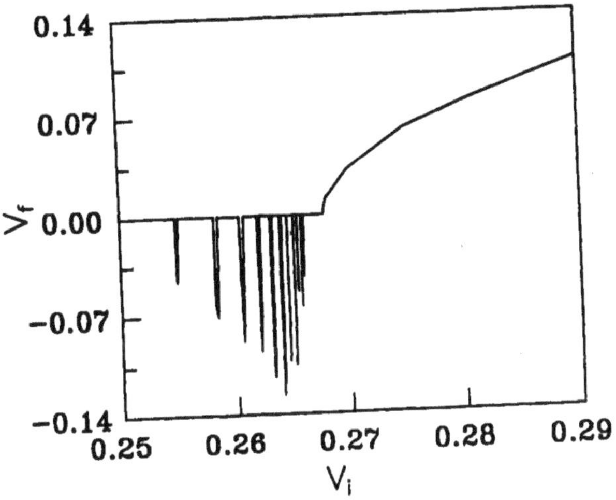

Figure 1. The final velocity of the SG kink as a function of its initial velocity at $\epsilon = 0.7$. Zero final velocity means that the kink is captured by the impurity.

- In quantitative analysis, we find that the time interval between the first and second interactions can be estimated by

$$T_{12}(V_i) = \frac{a}{\sqrt{V_c^2 - V_i^2}} + b, \tag{6}$$

where V_i is the kink initial velocity, a and b are two parameters, e.g., at $\epsilon = 0.7$ they are found to be $a = 3.31893$ and $b = 1.93$.

- The resonance effects can be explained by an energy exchange mechanism between the kink translation mode and the impurity mode. We have observed that the reflection of the kink is carried out in two steps, the first interaction excites the impurity mode and traps the kink, while the second interaction is just a reverse process, it will extinguish the impurity mode, retransferring enough energy back to the kink, and thus allowing it to escape. In this case, a resonance condition is satisfied

$$T_{12}(V) = nT_{im} + \tau, \tag{7}$$

where T_{im} is the period of the impurity mode oscillation, τ is an offset phase, and n is an integer.

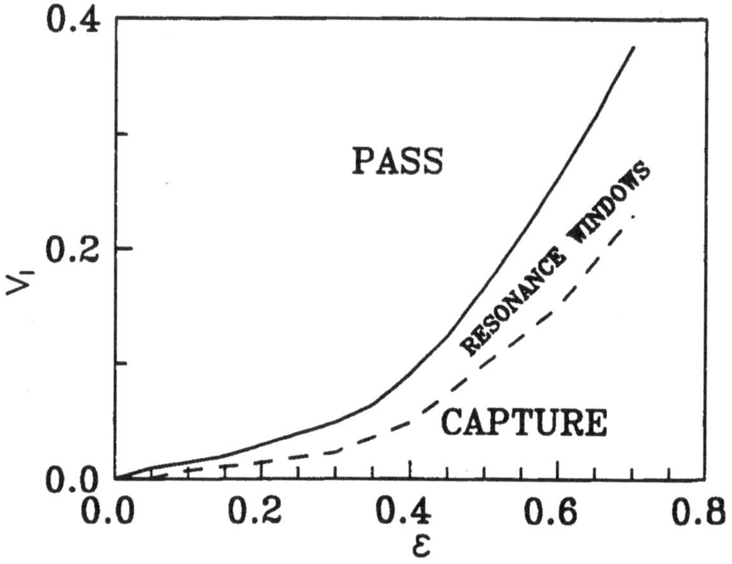

Figure 2. Three zones on the plane (ϵ, V_i) for the kink-impurity interactions in the SG model.

- Combining Eqs.(6) and (7), we may obtain a general formula to predict the centers of the resonance windows,

$$V_n^2 = V_c^2 - \frac{a^2}{(nT_{im} + \tau - b)^2}, \quad n = 2, 3, \ldots.. \tag{8}$$

  This formula can give very accurate predictions in comparison with the numerical results.

- The plane (ϵ, V_i) can be divided into three zones: transmission, capture and coexistence of capture and reflection (Fig. 2)

To study analytically the kink-impurity interactions in the SG model, we use a collective-coordinate method. The fundamental approximation involved in this approach is the replacement of the (infinitely) many degrees of freedom of a kink in a continuum field equation, by a (finite) number of "collective" variables. Such a reduction is usually helpful for one to extract the main physical mechanism underlying the observed nonlinear phenomenon.

By substituting the ansatz,

$$u = 4\tan^{-1}\exp[x - X(t)] + a(t)e^{-\epsilon|x|/2}, \tag{9}$$

into the Lagrangian of the system,

$$L = \int_{-\infty}^{\infty} dx \left[\frac{1}{2} u_t^2 - \frac{1}{2} u_x^2 - (1 - \epsilon\delta(x))(1 - \cos u) \right], \tag{10}$$

we may derive the following (reduced) effective Lagrangian,

$$L_{eff} = 4\dot{X}^2 + \frac{1}{\epsilon}(\dot{a}^2 - \Omega^2 a^2) - U(X) - aF(X), \tag{11}$$

where $U(X) = -2\epsilon/\cosh^2 X$, and $F(X) = -2\epsilon \tanh X/\cosh X$. The equations of motion for the two dynamical variables, X and a, are the following

$$\begin{cases} 8\ddot{X} + U'(X) + aF'(X) = 0, \\ \ddot{a} + \Omega^2 a + (\epsilon/2)F(X) = 0. \end{cases} \tag{12}$$

The system (12) describes a particle (kink) with the coordinate $X(t)$ and mass 8, scattered by an attractive potential $U(X)$ ($\epsilon > 0$), and "weakly" coupled with a harmonic oscillator $a(t)$ (the impurity mode). Here we say "weakly" because the coupling term $aF(X)$ is of order $O(\epsilon)$ and it decays exponentially. It has been shown that the dynamical model (12) can describe the main features of the kink-impurity interactions. In particular, it can be used to estimate the critical velocity of kink capture and to predict the resonance phenomena [4,5,6].

3 Kink-impurity interactions in the ϕ^4 model

Now let us consider the kink-impurity interactions in the ϕ^4 model [7]

$$u_{tt} - u_{xx} + [1 - \epsilon\delta(x)](u^3 - u) = 0, \tag{13}$$

Unlike the SG model, the pure ϕ^4 equation *is not integrable*, but it also admits kinks, that have an internal oscillating mode with the frequency $\omega_1 = \sqrt{3/2}$. As has been already shown [13], existence of the kink internal mode is very important for the resonance phenomena observed in the kink-antikink collisions.

In the presence of the impurity, the ϕ^4 equation (13) also has an impurity mode: $u(x,t) = \pm 1 + a_0 \cos(\Omega t + \theta) e^{-\epsilon|x|}$, where Ω and ϵ are connected by the dispersion relation $\Omega^2 = 2 - \epsilon^2$.

By numerical simulations we also observe a set of resonance windows in which the kink can be reflected by an attractive impurity (Fig.3). We find that both the kink internal mode and the impurity mode are excited during the scattering, and the resonance phenomena observed are due to an energy exchange between the kink translational mode, kink internal mode, and the impurity mode. Moreover, we observe a new interesting feature in the resonant interactions: some intermediate windows are missed, showing quasiresonance effects. Although a much smaller velocity step has

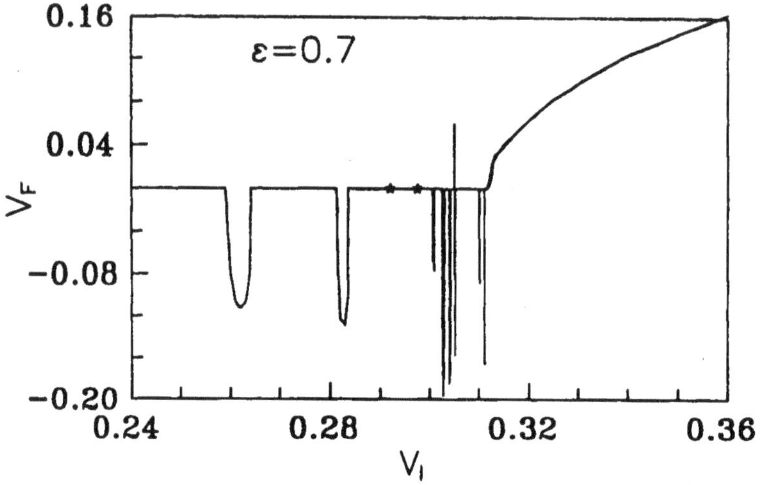

Figure 3. The same as in Fig.1 but for the ϕ^4 model (13) with $\epsilon = 0.7$. The two stars indicate the disappeared windows.

been used to scan these windows, we have not found real kink reflection. Instead we have only observed quasi-resonances, i.e., if the kink initial velocity is close to some values, the second interaction will reflect the kink at a larger distance from the impurity, but the kink cannot escape to infinity. Therefore, the resonance structures observed here, are quite different from those for the SG model in which the resonance windows come out one by one and they can be easily predicted by a simple formula. Furthermore, here we have observed a "three-bounce" resonance window in which the kink can escape after colliding with the impurity for three times. Since the impurity mode and the internal mode are excited in the scattering and they can be considered as small-amplitude oscillating states, for the collective-coordinate approach we propose the following ansatz:

$$u(x,t) = \tanh\left[\frac{x - X(t)}{\sqrt{2}}\right] + A(t)\eta(x - X(t)) + a(t)e^{-\epsilon|x|}, \tag{14}$$

where $\eta(x)$ is the kink internal (shape) mode, $A(t)$ being its amplitude, and $a(t)$ is the amplitude of the impurity mod. Substituting Eq.(14) into the system Lagrangian, in the lowest-order approximation we obtain the following (reduced) effective Lagrangian

$$\begin{aligned} L_{eff} = {} & \frac{M_k}{2}\dot{X}^2 - U(X) + \frac{1}{2}\dot{A}^2 - \frac{\omega_1^2}{2}A^2 - AF(X) \\ & + (1/\epsilon)[\frac{1}{2}\dot{a}^2 - \frac{\Omega^2}{2}a^2] - aG(X) - aAD(X), \end{aligned} \tag{15}$$

where $\omega_1 = \sqrt{3/2}$ and $\Omega = \sqrt{2-\epsilon^2}$ are the frequencies of the kink internal mode and the impurity mode, respectively, and $M_k = 2\sqrt{2}/3$ is the kink mass. The functions in (15) are

$$U(X) = -\frac{\epsilon}{4\cosh^4(X/\sqrt{2})}, \tag{16}$$

$$F(X) = (9/8)^{1/4}\frac{\epsilon\tanh^2(X/\sqrt{2})}{\cosh^3(X/\sqrt{2})}, \tag{17}$$

$$G(X) = -\frac{\epsilon\tanh(X/\sqrt{2})}{\cosh^2(X/\sqrt{2})}, \tag{18}$$

$$D(X) = (9/8)^{1/4}\frac{2\epsilon\tanh(X/\sqrt{2})}{\cosh(X/\sqrt{2})}. \tag{19}$$

The equations of motion derived from the Lagrangian (15) are

$$\left\{\begin{array}{l} M_k\ddot{X} + U'(X) + AF'(X) + aG'(X) + aAD'(X) = 0, \\ \ddot{A} + \omega_1^2 A + F(X) + aD(X) = 0, \\ (1/\epsilon)[\ddot{a} + \Omega^2 a] + G(X) + AD(X) = 0. \end{array}\right. \tag{20}$$

The system (20) describes a particle coupled with *two* oscillators, which correspond to the kink internal mode and the impurity mode, respectively. The two oscillators are also coupled through the term $aAD(X)$. The system (20) may be useful, e.g., to estimate analytically the amplitudes of the impurity mode and the kink internal mode that are excited during the kink scattering.

We have found that the system (20) can describe most of the features of the kink-impurity interactions observed in direct numerical simulations. To see this, we have simulated the system (20) under the suitable initial conditions corresponding to the problem of kink-impurity interactions. At $\epsilon = 0.5$, we find that there is a critical velocity $V_c \approx 0.26586$, above which the particle will pass the potential well inelastically, because it loses energy to excite the two oscillators. The final velocity of the particle satisfies the relation $V_f^2 = \alpha(V_i^2 - V_c^2)$, with $\alpha \approx 0.726$.

In the dynamical system (20) we observe the qualitatively similar structure of the resonance windows, e.g., some (intermediate) two-bounce resonance windows are missed showing quasiresonances. In fact, it is not difficult to understand why some two-bounce windows may disappear. Since the mechanism underlying the resonance is an energy exchange between the particle and the two oscillators, the resonance process can be devived into two steps: First, an incident particle loses its kinetic energy to excite the oscillators and get trapped by the potential well. Second, the particle comes back to the interaction region and has energy exchange again with the excited oscillators. Whether the particle can escape or not after the second interaction will depend on how much net energy the particle may obtain from the two oscillators. Because the frequencies of the oscillators are generally not commensurable, it may happen that even if the particle can restore energy from one, it may transfer energy to another. As a result, the particle cannot obtain enough kinetic energy to escape from the potential well. Therefore, instead of true resonances, only quasi-resonance effects are observed at certain initial velocities.

Since the system (20) is Hamiltonian, it easily allows higher-bounce (e.g., three-bounce) resonances, which means that a kink can be transmitted or reflected after colliding with the impurity for a number of times. However, in real interactions these resonances are strongly influenced by radiation, so they were not found in the SG model. Nevertheless, due to the interplay between several oscillating modes, such resonances are possible in the nonlinear systems supporting kinks with internal degrees of freedom, and we indeed observe a three-bounce resonance for the ϕ^4 model.

4 Multi-impurity kink scattering

Now we consider two point-like inhomogeneities in the system,

$$u_{tt} - u_{xx} + \{1 - \epsilon[\delta(x) + \delta(x - D)]\}F(u) = 0. \tag{21}$$

where $F(u) = \sin u$, or $u^3 - u$. A complete analysis of this model for the case $|\epsilon \ll 1$ has been carried out in Ref.[9,10]. Since the unperturbed SG equation is completely integrable, it is possible to use the perturbation theory based on the inverse scattering transform to compute the spectral energy density emitted by the kink upon collision with the two impurities. Integrating over the wave numbers corresponding to the waves scatterred backwards gives the reflected energy. In this way we can calculate the reflection coefficient.

We denote by R_1 and R_2 the reflection coefficients for one and two impurities, respectively. Then the following results may be obtained in the framework of perturbation theory (see Ref.[9] for detail):

$$\frac{R_2}{2R_1} \approx 1 + \frac{1}{(1 + D^2/\pi^2)^{1/4}} \cos\left[\frac{D}{v} - \frac{1}{4}\tan^{-1}\left(\frac{D}{\pi}\right)\right], \text{ for } v^2 \ll 1, \tag{22}$$

and

$$\frac{R_2}{2R_1} \approx 1 + 3\left(\frac{d}{\sinh d} + \frac{2\cosh d}{\sinh^2 d} - \frac{2d\cosh^2 d}{\sinh^3 d}\right), \text{ for } (1 - v^2) \ll 1, \tag{23}$$

where $d \equiv 2D/\sqrt{1 - v^2}$, and v is the kink velocity.

The above analytical results are shown in Fig.4 (solid lines). It is seen that fast kink do not show any signature of interference effects, while a slow one exhibits oscillations in the reflection coefficient as a function of the distance between the impurities. Also we notice that if $D \to 0$, then $R_2/2R_1 \to 2$; if $D \to \infty$, then $R_2/2R_1 \to 1$ (the scattering by the two impurities becomes independent and no interference effects arise).

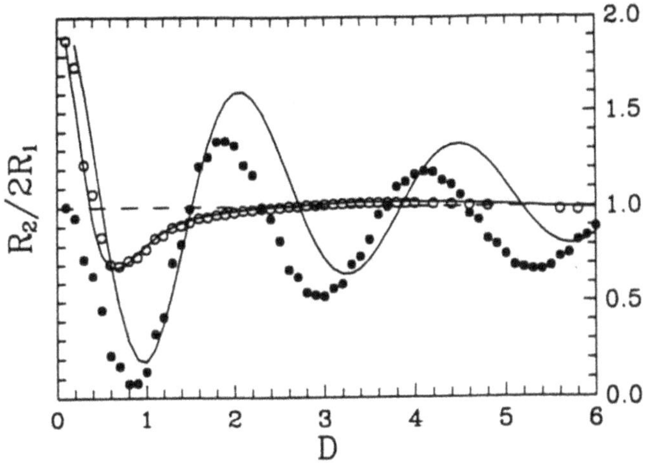

Figure 4. The ratio of the reflection coefficient of two impurities to that of one impurity. Lines are the analytical results given by Eqs. (22), (23). Circles are numerical results: empty for $v = 0.9$ kinks, and full for $v = 0.4$ kinks.

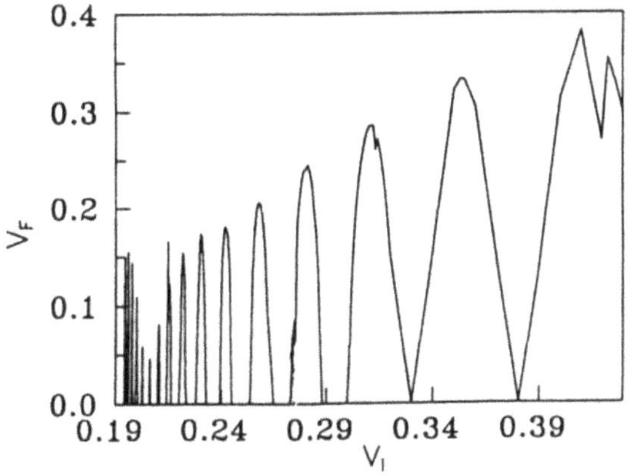

Figure 5. Final vs. initial kink velocity for the scattering by two (attractive) impurities in the ϕ^4 model ($\epsilon = 0.5$). Zero final velocity means that the kink is captured by one of the impurities.

As for the ϕ^4 model we distinguish two cases corresponding to repulsive and attractive impurities. The first case has been addressed in Ref.[9], where similar interference effects have been found numerically (in fact, they can also be explained by using a direct perturbation theory). In the case of attractive impurities, the problem has been studied in Ref. [10]. We found that, for a given large distance D, the final results of the kink scattering sensitively depend on the kink incoming velocity. In particular, we have observed a series of resonance windows where the kink can escape from the second impurity, due to the restoring of energy from its internal mode, which is excited by the collision with the first impurity (Fig.5).

The results about kink interactions with two impurities show that, if the impurities are separeted far away, the scattering process can be regarded as independent. Therefore, it is possible to treat the many impurity problem in the dilute limit assuming independent scattering. As an example, let's consider the problem in the framework of the SG model

$$u_{tt} - u_{xx} + [1 - \epsilon \sum_n \delta(x - a_n)] \sin u = 0. \tag{24}$$

where the distances $b_n = a_{n+1} - a_n$ are randomly distributed with the probability $p(b) = b_0 \exp(-b/b_0)$. If the mean distance between the impurities is large, we can treat the kink-impurity scatterings independently. This is done with the collective-coordinate method, in which a kink is considered as a classical particle and the δ-functions are substituted by equivalent sech—shaped potentials of width Δ which follow from the collective-coordinate formalism. The key point now is to describe the scattering in terms of the kink statistical characteristics, introduced by mapping of the spatial problem to the temporal one. This may be done considering statistics of the time intervals, τ_n when the kink travels between the nth and $(n+1)$th impurities, and the constant interval τ_0 between them during which the kink passes one impurity. In this way, the mean kink position, $X(t)$, as well as other statistical characteristics of the kink scattering can be calculated [8].

5 Conclusions

The resonant effects observed in the kink-impurity interactions in the SG and ϕ^4 models are analogous to the resonance phenomena in the kink-antikink collisions of some nonlinear Klein-Gordon equations [13]-[15]. The main physical mechanism of the resonances is an energy exchange between the kink translational motion and one, or several localized oscillating modes. In particular, if there is only one localized oscillating mode involved the resonance structures are regular, otherwise the resonance structures may be very complicated. Another important conclusion is that the collective-coordinate method is suitable for analyzing the kink-impurity interactions.

Our study may be extented in several directions: First, we may consider interaction of nontopological solitons with impurities, e.g., the breather-impurity collisions in the SG model. It is also interesting to investigate the resonant kink dynamics in more realistic dissipative systems such as long Josephson junctions. Another open question is whether it is possible to demonstrate the effects of "*true*" kink capture in the framework of a collective-coordinate system of finite (two or three) degrees of freedom that neglects radiation. Strictly speaking, this question cannot be answered by mere numerical simulations of the collective coordinate system since it can be performed only within a finite time interval. In fact, the collective coordinate systems arised in the present study can be considered as nearly-integrable ones since the coupling between the particle and the oscillators is small and localized. Therefore it would be useful to analyze more rigorously the mechanical model in the perspectives of chaos and fractals (see, e.g., some results for the kink-antikink collisions in Ref. [16]).

Finally, we would like to point out that, it is almost an open field the analysis of the numerical schemes for stochastic differential equations. Mathematical context of the above nonlinear wave equations with singular/random perturbations is a subject for future study.

One of us (Z.F.) acknowledges the Ministry of Education and Science of Spain for a fellowship. The work of Yu. K. has been supported by the Alexander von Humboldt-Stiftung. L. V. was partially supported by the Comision Interministerial de Ciencia y Tecnología of Spain under grant MAT90/0544.

References

[1] *"Disorder and Nonlinearity"*. Eds. A.R. Bishop, D.K. Campbell, and St. Pnevmatikos (Springer-Verlag, Berlin, 1989).

[2] A. Sánchez, L. Vázquez, *Int. J. Mod. Phys.* **B 5** (1991) 2825.

[3] S.A. Gredeskul and Yu. S. Kivshar, *Phys. Rep.* **216** (1992) 1.

[4] Yu.S. Kivshar, Zhang Fei, and L. Vázquez, *Phys. Rev. Lett.* **67** (1991) 1177.

[5] Zhang Fei, Yu.S. Kivshar, B.A. Malomed, and L. Vázquez, *Phys. Lett.* **A 159** (1991) 318.

[6] Zhang Fei, Yu.S. Kivshar, and L. Vázquez, *Phys. Rev.* **A 45** (1992) 6019.

[7] Zhang Fei, Yu.S. Kivshar, and L. Vázquez, *Phys. Rev.* **A 46** (1992) 5214.

[8] S.A. Gredeskul, Yu.S. Kivshar, L.K. Maslov, A. Sánchez, and L. Vázquez, *Phys. Rev.* **A 45** (1992) 8867.

[9] Yu.S. Kivshar, A. Sánchez, O. Chubykalo, A. M. Kosevich, and L. Vázquez, *J. Physics* **A 25** (1992) 5711.

[10] Zhang Fei, L. Vázquez, and Yu.S. Kivshar, *Phys. Lett. A*, (1992) submitted.

[11] Q. Li, St. Pnevmatikos, E.N. Economou, and C.M. Soukoulis, *Phys. Rev.* **B 37** (1988) 3534.

[12] Yu.S. Kivshar, S.A. Gredeskul, A. Sánchez, and L. Vázquez, *Phys. Rev. Lett.* **64** (1991) 1693.

[13] D.K. Campbell, J.F. Schonfeld, and C.A. Wingate, *Physica* **D 9** (1983) 1.

[14] M. Peyrard and D.K. Campbell, *Physica* **D 9**, 33 (1983); **D 18** (1986) 47.

[15] D.K.Campbell, M. Peyrard, and P. Sodano, *Physica* **D 19** (1986) 165.

[16] P. Anninos, S. Oliveira, and T.A. Matzner, *Phys. Rev.* **D 44** (1991) 1147.

FOURIER CHEBYSHEV APPROXIMATION FOR UNILATERALLY PERIODIC FLOW

GUO BEN-YU
Shanghai University of Science and Technology, Shanghai, China, 201800

ABSTRACT
A Fourier-Chebyshev spectral and pseudospectral schemes are proposed. A new filtering technique improves the stability and the accuracy. Error estimation and numerical results are presented.

1. INTRODUCTION

In studying boundary layer and other related problems, we have to consider unilaterally periodic problems (see [1-3]). There are several ways to solve such problems numerically. Murdok [1], Ingham [2], Moin, Kim [3], Macaraeg [4] and Guo Ben-yu, Xiong Yue-shan [5] proposed Fourier spectral (or pseudospectral)-finite difference schemes, which can be implemented easily. On the other hand, Canuto, Maday, Quarteroni [6], Guo Ben-yu, Cao Wei-ming [7], and Guo Ben-yu, Ma He-ping [8] provided Fourier spectral (or pseudospectra)-finite element schemes, which are suitable for complex geometry. But the accuracies of the above methods are still limited duo to the finite difference and finite element approximations, even if the genuine solutions are very smooth. In this paper, we develop other mixed methods by using Fourier spectral (or pseudospectral) approximation in the periodic direction and Chebyshev spectral (or pseudospectral) approximation in the other directions. Such methods are as precise as fully spectral (or pseudospectral) method. In particular, the Fourier-Chebyshev pseudospectral method can deal with nonlinear terms more efficiently and saves a lot of work.

2. THE SCHEMES

Let $I_x = (-1,1), I_y = (0,2\pi)$ and $\Omega = I_x \times I_y$. We denote the vorticity, stream function and kinetic viscosity by $\xi(x,y,t), \psi(x,y,t)$ and ν respectively, $f_j(x,y,t)$ and $\xi_0(x,y,t)$ are given functions. We consider the two-dimensional vorticity equation as follows

$$\begin{cases} \frac{\partial \xi}{\partial t} + J(\xi,\psi) - \nu\nabla^2\xi = f_1, & \text{in } \Omega \times (0,T], \\ -\nabla^2\psi = \xi + f_2, & \text{in } \Omega \times [0,T], \\ \xi(x,y,0) = \xi_0(x,y), & \text{in } \overline{\Omega} \end{cases} \tag{1}$$

where

$$J(\xi,\psi) = \frac{\partial \xi}{\partial x}\frac{\partial \psi}{\partial y} - \frac{\partial \xi}{\partial y}\frac{\partial \psi}{\partial x}.$$

Assume that all functions in (1) have the period 2π for the variable y, but satisfy non-periodic boundary conditions for $|x| = 1$. For simplicity, suppose that $\xi(\pm 1, y, t) = \psi(\pm 1, y, t) = 0$ for all $y \in I_y$ and $t \le T$.

Let M and N be any positive integers, and c_1, c_2 be some positive constants such that

$$c_1 N \le M \le c_2 N. \tag{2}$$

Let Q_M denote the set of algebraic polynomials of degree less or equal M, and define

$$V_M(I_x) = \{v(x) \in Q_M / v(-1) = v(1) = 0\}.$$

Let $V_N(I_y)$ be the set of all real trigonometric polynomials defined on I_y, with the period 2π and the degree less or equal N. Furthermore $V_{M,N}(\Omega) = V_M(I_x) \times V_N(I_y)$.

Let $\omega(x) = (1 - x^2)^{-\frac{1}{2}}$ and define the space

$$L^2_\omega(\Omega) = \{\nu \text{ is measurable}/\|v\|_\omega < \infty\}$$

equipped with the following inner product and norm

$$(u,v)_\omega = \frac{1}{2\pi}\int_\Omega u(x,y)v(x,y)w(x)dxdy, \quad \|v\|_\omega = (v,v)^{\frac{1}{2}}_\omega.$$

Let $P_{M,N} : L^2_\omega(\Omega) \to V_{M,N}$ be the orthogonal projection, i.e., for any $v \in L^2_\omega(\Omega)$,

$$(P_{M,N}v - v, u)_\omega = 0, \quad \forall u \in V_{M,N}.$$

Let τ be the step of time t, and define

$$S_\tau = \{t = k\tau / 1 \le k \le [\frac{T}{\tau}]\},$$

$$v_{\hat{t}}(t) = \frac{1}{2\tau}(v(t+\tau) - v(t-\tau)).$$

Now, let η and φ be the approximations to ξ and ψ respectively. The Fourier-Chybeshev spectral scheme for (1) is to find $(\eta,\varphi) \in V_{M,N}(\Omega) \times V_{M,N}(\Omega)$ such that

$$\begin{cases} \eta_{\hat{t}}(t) + P_{M,N}J(\eta(t),\varphi(t)) - \frac{\nu}{2}\nabla^2(\eta(t+\tau) + \eta(t-\tau)) = P_{M,N}f_1, & \text{for } t \in S_\tau, \\ -\nabla^2\varphi(t) = \xi(t) + P_{M,N}f_2, & \text{for } t \in S_\tau \bigcup\{0\}, \\ \eta(\tau) = P_{M,N}(\xi_0 + \tau\frac{\partial \xi}{\partial t}(0)), & \\ \eta(0) = P_{M,N}\xi_0, & \end{cases} \tag{3}$$

where

$$\frac{\partial \xi}{\partial t}(0) = \nu \nabla^2 \xi_0 - J(\xi_0, \psi_0) + f(0).$$

Since FFT is valid for both Fourier and Chebyshev expansions, we can evaluate $\eta(t)$ and $\varphi(t)$ simply. But it is not easy to deal with the nonlinear terms. To remedy this deficiency, we prefer to adopt pseudospectral method. Numerical experiments show that pseudospectral method is not so stable as spectral one, due to the aliasing of interaction. We also know that the Fourier summations are not convergent for functions in $L^1(\Omega)$ generally, such as shock wave. There are two main ways to remove these deficiencies. The first was proposed by Kuo Pen-yu [9], based on Bochner summation. The second was given by Woodward, Collela and Vandeven, see [10,11]. We generalize the first one to Chebyshev approximation. For the mixed method, let $\gamma_1, \gamma_2 \geq 1$ and $R = R(\gamma_1, \gamma_2)$ be the filtering operator. It means that if

$$v = \sum_{j=0}^{M} \sum_{|l| \leq N} a_{jl} T_j(x) e^{ily}.$$

T_j being the Chebyshev polynomial of degree j, then

$$Rv = \sum_{j=0}^{M} \sum_{|l| \leq N} a_{jl} (1 - |\frac{j}{M}|^{\gamma_1})(1 - |\frac{l}{N}|^{\gamma_2}) T_j(x) e^{ily}.$$

Let $\{x_j\}$ and $\{w_j\}$ be the nodes and weights of Gauss-Labatto integration. Also let $h = \frac{2\pi}{2N+1}$ be the mesh size for the variable y. Define

$$\Omega_{M,N} = \{(x,y)/x = x_j, y = lh, 0 \leq j \leq M, 0 \leq l \leq 2N\}.$$

We by P_c denote the interpolation from $C(\overline{\Omega})$ to $V_{M,N}(\Omega)$, i.e.,

$$P_c v(x,y) = v(x,y), \quad \text{for } (x,y) \in \Omega_{M,N}.$$

The another key point is to approximate the nonlinear terms reasonably. To do this, define

$$J_c(u,v) = \frac{\partial}{\partial x}[P_c(u\frac{\partial v}{\partial y})] - \frac{\partial}{\partial y}[P_c(u\frac{\partial v}{\partial x})].$$

Let η_c and φ_c be the approximations to ξ and ψ respectively. The Fourier-Chebyshev pseudospectral scheme for (1) is to fined $(\eta_c, \varphi_c) \in V_{M,N} \times V_{M,N}$ such that

$$\begin{cases} \eta_{c,\hat{t}}(t) + RJ_c(R\eta_c(t), R\varphi_c(t)) - \frac{\nu}{2}\nabla^2(\eta_c(t+\tau) + \eta_c(t-\eta)) = P_c f_1, & t \in S_\tau, \\ -\nabla^2 \varphi_c = \eta_c + P_c f_2, & t \in S_\tau \bigcup \{0\}, \\ \eta_c(\tau) = P_{M,N}(\xi_0 + \tau \frac{\partial \xi}{\partial t}(0)), \\ \eta_c(0) = P_{M,N}\xi_0. \end{cases} \quad (4)$$

3. ERROR ESTIMATION

We first introduce some Cobolev spaces of functions defined on I_x with the weight $\omega(x)$. For any integer $r \geq 0$, set

$$|v|_{r,w,I_x} = \left(\int_{I_x} \left(\frac{d^r v}{dx^r}\right)^2 w dx\right)^{\frac{1}{2}}, \quad \|v\|_{r,w,I_x} = \left(\sum_{j=0}^{r} |v|_{j,w,I_x}^2\right)^{\frac{1}{2}}$$

and

$$H_w^r(I_x) = \{v / \|v\|_{r,w,I_x} < \infty\}.$$

In particular $H_w^0(I_x) = L_w^2(I_x)$ and $\|v\|_{0,w,I_x} = \|v\|_{w,I_x}$. For any real $r > 0, H_w^r(I_x)$ is defined by the complex interpolation between the spaces $H_w^{[r]}(I_x)$ and $H_w^{[r+1]}(I_x)$. Moreover

$$C_0^\infty(I_x) = \{v/v \text{ is infinitely differentiable, and } v(-1) = v(1) = 0\}.$$

Denote by $H_{0,w}^r(I_x)$ the Closure of $C_0^\infty(I_x)$ in $H_w^r(I_x)$.

Next, let B a Banach space with the norm $\|\cdot\|_B$, and Γ be an interval in R. Define

$$L^2(\Gamma, B) = \{v(z) : \Gamma \to B/v \text{ is strongly measurable and } \|v\|_{L^2(\Gamma,B)} < \infty\}$$

where

$$\|v\|_{L^2(\Gamma,B)} = \left(\int_\Gamma \|v(z)\|_B^2 dz\right)^{\frac{1}{2}}.$$

For any integer $s \geq 0$, let

$$H^s(\Gamma, B) = \{v(z) \in L^2(\Gamma, B)/\|v\|_{H^s(\Gamma,B)} < \infty\}$$

equipped with the norm

$$\|v\|_{H^s(\Gamma,B)} = \left(\sum_{k=0}^{s} \left\|\frac{\partial^k v}{\partial z^k}\right\|_B^2\right)^{\frac{1}{2}}.$$

For real $s > 0$, we can define $H^s(\Gamma, B)$ by the complex interpolation of spaces as before. Similarly we can define the space $C^s(\Gamma, B)$.

For simplifying the statements, we also introduce some non-isotropic spaces. Let

$$H_w^{r,s}(\Omega) = L^2(I_y, H_w^r(I_x)) \bigcap H^s(I_y, L_w^2(I_x))$$

with the norm

$$\|v\|_{H_w^{r,s}(\Omega)} = \left(\|v\|_{L^2(I_y,H_w^r(I_x))}^2 + \|v\|_{H^s(I_y,L_w^2(I_x))}^2\right)^{\frac{1}{2}}.$$

Moreover define

$$M_w^{r,s}(\Omega) = H_w^{r,s}(\Omega)\bigcap H^1(I_y, H_w^{r-1}(I_x))\bigcap H^{s-1}(I_y, H_w^1(I_x)),$$
$$X_w^{r,s}(\Omega) = H_w^{2,2}(\Omega)\bigcap H^s(I_y, H_w^r(I_x))\bigcap H^{s+1}(I_y, H_w^{r-1}(I_x)).$$

The meaning of norms $\|v\|_{M_w^{r,s}(\Omega)}$ and $\|v\|_{X_w^{r,s}(\Omega)}$ are clear. Also let $C_{0,p}^{\infty}(\Omega)$ be the set of all infinitely differentiable functions defined on $\overline{\Omega}$, which vanish at $|x| = 1$ and have the period 2π for y. Denote by $H_{0,p,w(\Omega)}^{r,s}$ the closure of $C_{0,p}^{\infty}(\Omega)$ in $H_w^{r,s}(\Omega)$, etc..

Let d be some positive constant depending only on ν and the norms of ξ and ψ in the spaces mentioned in following theorems.

THEOREM 1. *Assume that*

(i) $\xi, \psi \in C(0,T; M_{0,p,w}^{r,s}(\Omega)\bigcap X_w^{\alpha,\beta}(\Omega)), r, s \geq 1, \alpha > \frac{5}{2}, \beta > \frac{1}{2}$,

(ii) $\frac{\partial\xi}{\partial t} \in C(0,T; M_w^{r,s}(\Omega)), \frac{\partial^2\xi}{\partial t^2} \in L^2(0,T; H_w^{1,1}(\Omega)), \frac{\partial^3\xi}{\partial t^3} \in L^2(0,T; L_w^2(\Omega)), f_1, f_2 \in C(0,T; H_w^{r,s}(\Omega))$,

(iii) (2) holds and $\tau = O(M^{-\frac{1}{4}})$.

Then for all $t \leq T$,

$$\|\xi(t) - \eta(t)\|_w \leq d_1(\tau^2 + M^{-r} + N^{-s}).$$

THEOREM 2. *Let conditions (i), (ii) of Theorem 1 hold, and*

$$\xi \in C(0,T; M_{0,p,w}^{r+1,s+1}(\Omega)), \psi \in C(0,T; M_{0,p,w}^{r+1,s+1}(\Omega)\bigcap X_w^{\alpha,\beta}(\Omega)), r, s, \beta \geq 1, \alpha > \frac{2}{5}.$$

Then for all $t \leq T$,

$$\|\xi(t) - \eta_c(t)\|_w \leq d(\tau^2 + M^{-r} + N^{-s}).$$

The proof of Theormes can be found in [12,13].

4. NUMERICAL RESULTS

For describing the errors, let

$$E(\tilde{\xi}, t) = \sqrt{\frac{\sum_{(x,y)\in\Omega_{M,N}} |\xi(x,y,t) - \eta^*(x,y,t)|^2}{\sum_{(x,y)\in\Omega_{M,N}} |\xi(x,y,t)^2|}}$$

where η^* is the numerical solution. We take the test functions

$$\xi = \frac{2}{5}e^{\frac{1}{2}t}(x^2-1)(x^2-8)\sin 2y - \frac{6}{5}Ae^{\frac{1}{2}t}(x^2-1),$$
$$\psi = \frac{1}{10}e^{\frac{1}{2}t}(x^2-1)(x^2-5)(\sin 2y + A).$$

We use schemes (3) and (4) to solve (1) with $\tau = 0.01, M = N = 4, M^* = 10, \gamma_1 = \gamma_2 = \infty$. For comparison, we consider also the Fourier spectral-linear finite element scheme (FSFE) and Fourier pseudospectral-linear finite element scheme (FPSFE), in which I_x is uniformly partitioned with the mesh size $h^* = \frac{2}{M^*}$. The numerical results show that (3) and (4) give much better results than FSFE and FPSFE methods. They provide the approximate solutions with very high accarcy, even if M and N are small (see Table 1 and Table 2).

Table 1. $\nu = 0.001, A = 0.$

t=5	scheme(3)	scheme(4)	FSFE	FPSFE
$E(\tilde{\xi},t)$	0.3027-4	0.6735-4	0.7188-2	0.2053-3

Table 2. $\nu = 0.0001, A = 0.$

t=5	scheme(3)	scheme(4)	FSFE	FPSFE
$E(\tilde{\xi},t)$	0.9165-4	0.9477-4	0.1284-1	0.8589-3

Table 3 shows the numerical results of scheme (4) with $\tau = 0.005, M = N = 4$ and the different choice of γ_1 and γ_2. Clearly the filtering operator $R(\gamma_1, \gamma_2)$ improves the stability and raises the accuracy.

Table 3. $\nu = 0.001, A = 0.01.$

$E(\tilde{\xi},t)$	$\gamma_1 = \gamma_2 = \infty$	$\gamma_1 = \gamma_2 = 20$	$\gamma_1 = \gamma_2 = 10$
$t = 0.5$	0.6105-4	0.5735-4	0.4224-4
$t = 1.5$	0.2213-3	0.2068-3	0.1390-3
$t = 2.5$	0.6793-3	0.6235-3	0.3745-3

References

[1] J. W. Murdok, *AIAA*, **86-0434** (1986) 1.

[2] D. B. Ingham, *Proc. R. Soc. London*, **A402** (1985) 109.

[3] P. Moin and J. Kim, *J. Fluid Mech.* **118** (1982) 341.

[4] M. G. Macaraeg, *J. Comp. Phys.* **62** (1982) 297.

[5] Guo Ben-yu and Xiong Yue-shan, *J. Comp. Phys.* **84** (1989) 259.

[6] C. Canuto, Y. Maday and A. Quarteroni, *Numer. Math.* **44** (1984) 201.

[7] Guo Ben-yu and Cao Wei-ming, *HMS Bulletin*, **32** (1991) 83.

[8] Guo Ben-yu and Ma He-ping, *SIAM J. Numer. Anal.* **28** (1991) 113.

[9] Kuo Pen-yu, *J. Comp. Math.* **1** (1983) 353.

[10] P. Woodward and P. Collela, *J. Comp. Phys.* **54** (1984) 115.

[11] H. Vandeven, *CNRS F56, Centre de Math. Appl., Ecole Polytechinique*, 1987.

[12] Guo Ben-yu, Ma He-ping, Cao Wei-ming and Huang Hui, *J. Comp. Phys.* **101** (1992), 207.

[13] Guo Ben-yu and Li Jian, Fourier-Chebyshev pseudospectral method for two-dimensional vorticity equation (unpublished).

On the Existence of a System of Nonlinear Partial Differential Equations with Singular Initial Data

Huang Dun
Peking University, Institute of Mathematics

We consider the gasdynamic equations in Eulerian coordinates x, t. The unknown functions are velocity u, pressure p and density ρ.

$$\frac{\partial u}{\partial t}+u\frac{\partial u}{\partial x}+\frac{1\partial p}{\rho\partial x}=0, \qquad \frac{\partial \rho}{\partial t}+\frac{\partial \rho u}{\partial x}=0, \tag{1.1}$$

$$(\frac{\partial}{\partial t}+u\frac{\partial}{\partial x})\frac{p}{\rho^{\gamma}}=0, \qquad \gamma=\text{const}>1, \text{say } 1.4$$

The boundary condition is $u(0,t)=0$ at $x=0$. (1.2)

The initial values are power functions with negative exponents, so **singular** at $x=0$.

$$\begin{aligned} p(x,0)&=\mathrm{L}\mathrm{A}x^{\alpha}, && \alpha<0. \\ \rho(x,0)&=\mathrm{N}\mathrm{B}x^{\beta}, && \beta<0, \quad \alpha-\beta<0 \\ u(x,0)&=M\sqrt{\frac{A}{B}}x^{(\alpha-\beta)/2}, && M\neq 0,\ n\neq 0. \end{aligned} \tag{1.3}$$

L, N, M, α, β are dimensionless constants and A, B are dimensional constants to give initial functions appropriate dimensions of pressure, density and velocity.

This system of equations is so called selfsimilar i.e. with the introduction of a combination of x and t, say a new independent variable λ

$$\lambda=\sqrt{\frac{A}{B}}tx^{\frac{\alpha-\beta}{2}-1}\geq 0 \tag{1.4}$$

we come to a system of ordinary differential equations, which has the following form:

$$\frac{dZ}{dV}=\frac{Z}{(V-q)}\{[(\gamma+1)V-2]\frac{(V-q)^2-Z}{\omega}-(\gamma-1)\} \tag{1.5}$$

$$q\frac{dln\lambda}{dV}=\frac{(V-q)^2-Z}{\omega} \tag{1.6}$$

$$(V-q)\frac{dlnR}{dln\lambda}=q\frac{\omega}{Z-(V-q)^2}+q(\beta+1)V \tag{1.7}$$

where

$$\omega=V(V-1)(V-q)+Z(K-V),\ q=\frac{2}{2-(\alpha-\beta)},\ K=-\frac{\alpha}{\gamma}q>0 \tag{1.8}$$

The unknown functions are now $V(\lambda), P(\lambda)$ and $R(\lambda)$ instead of u, p, and ρ.

$$u(x,t)=\frac{x}{t}V(\lambda), \qquad p(x,t)=Bx^{\beta+2}t^{-2}P(\lambda),$$

$$\rho(x,t) = Bx^{\beta}R(\lambda), \qquad Z(\lambda) = \gamma\frac{P(\lambda)}{R(\lambda)} > 0 \tag{1.9}$$

In the phase plane (V, Z) equation (1.5) has nine singularities at: $O(0,0), A(q,0), B(1,0)$,

$$C(K,\infty),\ D(q,\infty)\ ,E(\frac{2}{\gamma+1}, \frac{2(\gamma-1)[(\gamma+1)q-2]}{(\gamma+1)^2[2-(\gamma+1)K]}), F(\frac{qK}{q+K-1}, [\frac{q(q-1)}{q+K-1}]^2),$$

$G(\pm\infty,\infty)$ and $H(\pm\infty,0)$.

The field of lines in the (V, Z) plane, which are integrals of equation (1.5), are dependent upon three parameters γ, q, K, or α, β, γ.

Initial condition requires that as $t \to 0, \lambda \to 0$, and (1.9) tend to (1.3). This means that the required solution should start from the origin $V = 0, Z = 0$ in the (V, Z) plane, and has here the following asymptotic expansions:

$$V = M\lambda + V_a\lambda^2 + \cdots, \qquad P = L\lambda^2 + p_a\lambda^3 + \cdots,$$
$$R = N + R_a\lambda + \cdots, \qquad \lambda = -\frac{1}{M}V + \cdots. \tag{1.10}$$

For positive time $t > 0$, the origin corresponds to $x \to \infty$. We should seek the solution, which corresponds to the case , for which λ can vary continuously and monotonically from 0 to ∞, a solution with physical meaning for the whole domain $t > 0$ and $x > 0$. But equation(1.6) shows, that λ reaches its local maximum on the parabola $Z = (V - q)^2$. So the integration curve starting from origin before reaching the parabola should jump to another curve to ensure monotonicity of λ. This requires a discontinuous solution of V and Z. That is the solution jumps from one integration curve to another(Fig.1). We seek a solution which corresponds to shock wave reflection at $x = 0$, and the following shock wave jump conditions should also be imposed.

$$V_2 - q = (V_1 - q)[1 + \frac{2}{\gamma+1}\frac{Z_1 - (V_1-q)^2}{(V_1-q)^2}]$$
$$Z_2 = (\frac{\gamma-1}{\gamma+1})^2 \cdot \frac{1}{(V_1-q)^2}[(V_1-q)^2 + \frac{2Z_1}{\gamma-1}][\frac{2\gamma}{\gamma-1}(V_1-q)^2 - Z_1] \tag{1.11}$$

where the points $P_1(V_1, Z_1)$ and $P_2(V_2, Z_2)$ correspond to the points just in front of and just behind the reflection shock wave.

With detailed analysis of the nine singularities in the (V, Z) plane, especially two focus points O, E, at which $\lambda \to 0$, and saddle point,C, at which $\lambda \to \infty$, we proved the following theorem.

Theorem **A sufficient condition for the existence of global solution for the initial boundary value problem(1.1),(1.2) and (1.3) which corresponds to a shock wave reflection is:**

$$\alpha < \beta < 0; \quad (\frac{1}{2-\alpha+\beta} - \frac{1}{\gamma+1})(\frac{\alpha}{\gamma}\frac{1}{2-\alpha+\beta} + \frac{1}{\gamma+1}) > \frac{1}{4\gamma}(\frac{\gamma-1}{\gamma+1})^2. \tag{2}$$

This theorem helped us in selecting proper singular initial values.

The meaning of (2) in the theorem is just the requirements that the singular point E of fovus type, at which $\lambda \to 0$, should locate above the curve,

$$Z = (V - q)^2 \tag{2.1}$$

and lower than the curve

$$Z = \frac{2\gamma}{\gamma - 1}(V - q)^2 \qquad (2.2)$$

By shock wave relation (1.11) the points on V axis ($Z = 0$) jump to corresponding points on the curve (2.2), and points on (2.1) to itself.

The requirement that E locates above than (2.1) is satisfied automatically by the inequalities in the initial condition, posed by the motivation to simulate unsteady blast wave reflection, (1.3): $\alpha - \beta < 0$(so $0 < q < 1$), and $\alpha < 0$(so $K > 0$).

When the focus E lies on the curve (2.2), (1.11) gives

$$(q - \frac{2}{\gamma+1})(-K + \frac{2}{\gamma+1}) = \frac{1}{\gamma}(\frac{\gamma-1}{\gamma+1})^2 \qquad (2.3)$$

which by the notation (1.8), after simple manipulations, gives the second inequality of (2).

The requirement that E locates between curves (2.1) and (2.2)follows from the analysis of the topology of all the possible integration curves on (V, Z) plane.

These curves start from one of the nine singular points to another. The topology can be classified further into three cases. Fig.a shows the case $K > \frac{2}{\gamma-1}(1-q)$, or $\alpha - \gamma\beta > 0$. Fig.b shows the most important case for the modelling of the blast wave reflection from the plane $x = 0$. For this case $K < \frac{2}{\gamma-1}(1-q)$, or $\alpha - \gamma\beta < 0$, say the entropy of the gas initially decreases as x increases. Fig.c corresponds to the case $K = \frac{2}{\gamma-1}(1-q)$, or $\alpha - \gamma\beta = 0$, giving by the initial condition, p/ρ^γ to be independent of x, or initially homoentropic flow.

In the Fig.a to c, the arrows show the direction of increasing λ determined by equation (1.6). As stated above, λ should tend monotonically to ∞. The only singular point with λ infinity is the saddle point C. There is a curve connecting $C(K, \infty)$ with E(value of V at E being $2/(\gamma+1)$) From E to C the value of λ changes monotonically from 0 to ∞. This can not be the whole solution, since it does not contain the origin $V = Z = 0$, the initial data. Fig. 1 shows the appropriate integration curve starting from 0 to a point P_1. It jumps to P_2 with equal value of λ. Form P_2 the upper half of the curve EC serves as the remaining part of the required solution. The solution can be obtained easily by numerical integration of ordinary differential equations, which takes only a few seconds. During numerical integration the following asymptotic expansion for the saddle point C ($V = K, Z \to \infty$) should be used

$$\frac{1}{Z} = \frac{(q-K) + (2 - K - \gamma K)}{K(1-K)(q-K)^2}(V - K), \quad ln\lambda = -\frac{1}{q}\cdot\frac{q-K}{2-K-\gamma K}ln(V-K)$$

ACKNOWLEDGEMENTS

This paper is sponsored by the National Natural Science Foundation, and the National Committee of Science & Technology.

REFERENCES

Sedov L. I., The Method of Simularity and Dimensional Analyaia in Mechanics, Moscow.

Fig. a $k > \frac{2}{\gamma - 1}(1 - q)$ $(\alpha - \gamma\beta > 0)$

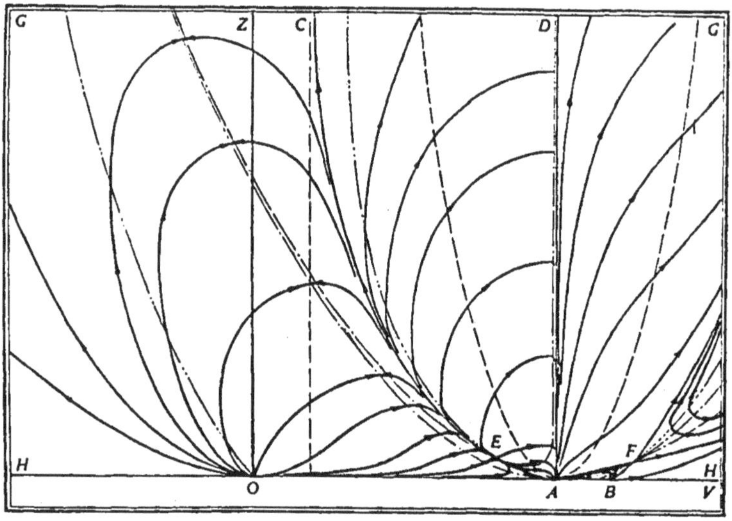

Fig. b. $k < \frac{2}{\gamma - 1}(1 - q)$ $(\alpha - \gamma\beta < 0,\ q + \kappa - 1 > 0)$

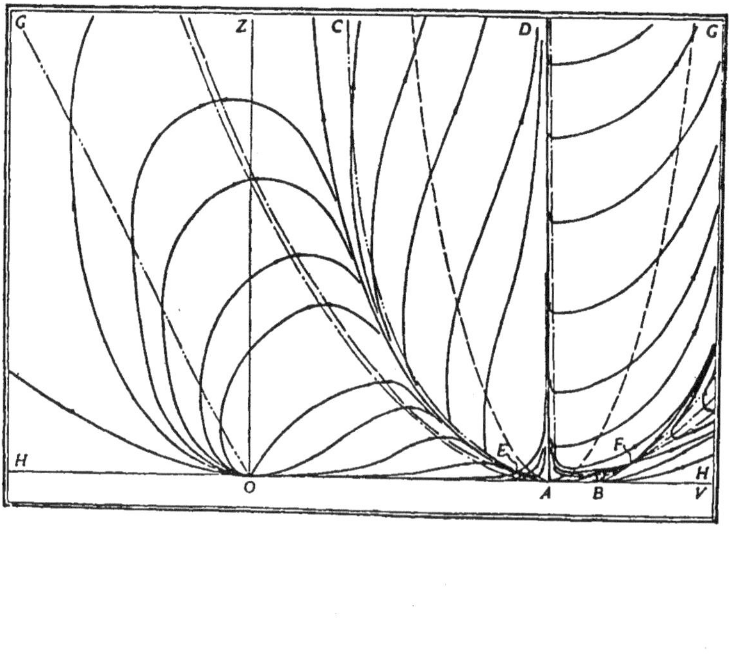

Fig. C $k = \frac{2}{\gamma - 1}(1 - q)$ $(\alpha - \gamma\beta = 0)$ Homoentropic flow

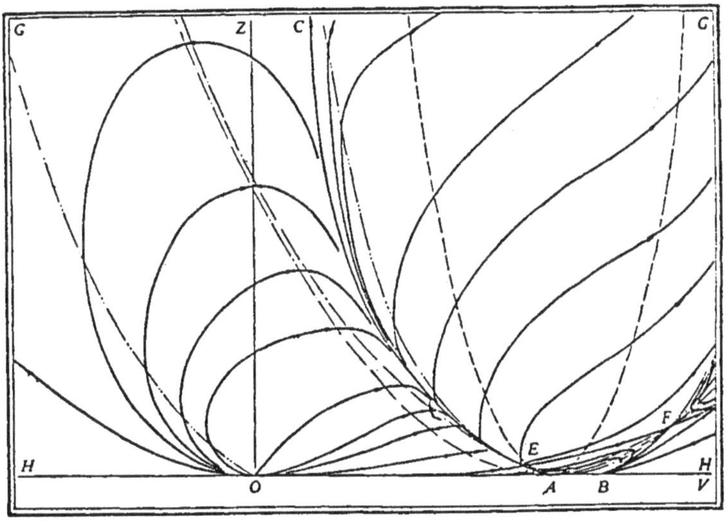

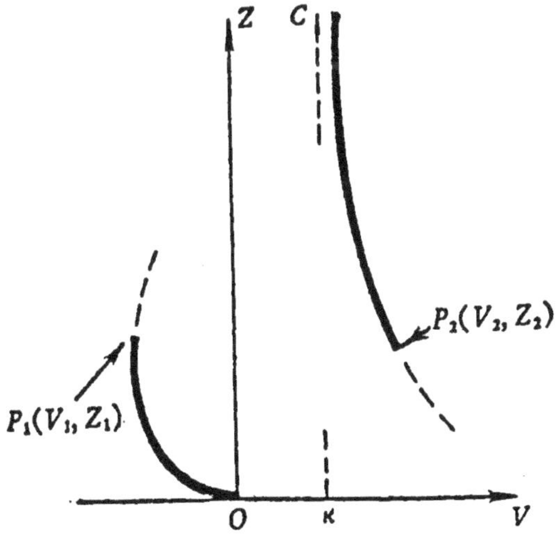

Fig.1

Reduced Stability Analysis for Magnetohydrodynamic Equilibria

TAKASHI KAKO

Department of Computer Science and Information Mathematics
The University of Electro-Communications (UEC)
Chofu-shi, Tokyo 182, JAPAN

Abstract. In the plasma confinement experiment for the controlled thermonuclear fusion research, the stability of a given equilibrium is one of the main concerns. We shall give an abstract formulation of the reduction process for the linearized stability analysis which reduces an analysis for 7×7 component operator into the one for a single operator. Some numerical difficulty related to this problem will also be disccussed.

1. Introduction

The non-dissipative classical motion of plasma is described by the following ideal magnetohydrodynamic system of equations which consists of the Euler equation and the reduced Maxwell equation with interaction terms given by the Lorenz force and Ohm's low:

$$\frac{\partial \rho}{\partial t} + \mathrm{div}(\rho v) = 0 \qquad \text{(conservation of mass)}$$

$$\rho \frac{Dv}{Dt} = -\mathrm{grad} P + j \times B \qquad \text{(equation of motion)}$$

$$E + v \times B = 0 \qquad \text{(Ohm's law)}$$

$$\frac{D}{Dt}(P\rho^{-\gamma}) = 0 \qquad \text{(adiabatic condition)}$$

$$\left.\begin{aligned} &\frac{\partial B}{\partial t} = -\mathrm{rot} E \\ &\mathrm{div} B = 0 \\ &j = \mathrm{rot} B \end{aligned}\right\} \qquad \text{(Maxwell equation)} \tag{1.1}$$

with

$$\frac{D}{Dt} = \frac{\partial}{\partial t} + v \cdot \mathrm{grad} \qquad \text{(Lagrangian derivative),}$$

where ρ, v, P, j, B and E are the density, the velocity, the pressure, the electric current , the magnetic field and the electric field respectively. The specific heat ratio γ is a constant. The equilibrium of the above system is a set of time independent solutions with zero velocity: $v_0 \equiv 0$:

$$\begin{aligned} \mathrm{grad} P_0 &= j_0 \times B_0 \\ j_0 &= \mathrm{rot} B_0 \\ \mathrm{div} B_0 &= 0. \end{aligned} \tag{1.2}$$

The linearized MHD equation around the equilibrium is given as follows.

$$\frac{\partial}{\partial t}\begin{pmatrix} Q \\ \rho_0 v \\ p \end{pmatrix} = \begin{pmatrix} 0 & -\mathrm{rot}(B_0 \times & 0 \\ -B_0 \times \mathrm{rot} + j_0 \times & 0 & -\mathrm{grad} \\ 0 & -\gamma P_0 \mathrm{div} - (\mathrm{grad} P_0)\cdot & 0 \end{pmatrix} \begin{pmatrix} Q \\ v \\ p \end{pmatrix}$$
$$\mathrm{div} Q = 0. \tag{1.3}$$

We call the 7×7 component differential operator which appears in the righ-hand side of the above equation the force operator F. The linearized stability analysis is a way to give a practical criterion if a given equilibrium is stable or not. We say that the equilibrium is stable (unstable) when the spectrum of the force operator is (is not) included in the left half complex plane. Eliminating Q and p, we can reduce the equation (3) to the equation for v :

$$\begin{aligned} \rho_0 \frac{\partial^2 v}{\partial t^2} = -Kv \equiv \mathrm{grad}\{\gamma P_0(\mathrm{div}\, v) + (\mathrm{grad}\, P_0) \cdot v\} \\ + B_0 \times \mathrm{rot}(\mathrm{rot}(B_0 \times v)) - (\mathrm{rot}\, B_0) \times \mathrm{rot}(B_0 \times v). \end{aligned} \tag{1.4}$$

The equilibrium is stable if the spectrum of K is a subset of a non-negative real semi-axis. The spectrum of F is a subset of an imaginary axis in this case. For a general reference on the spectral analysis for MHD operarors, see Lifschitz [4].

2. Reduced energy form and reduced stability analysis

In this section, we shall give some general ideas how to reduce further the stability analysis. In some cases with a practical importance, the operator K in section 1 can be Fourier decomposed and takes the form:

$$K = \begin{pmatrix} A & B \\ B^* & C \end{pmatrix} \tag{2.1}$$

where A, B, B^* and C are some differential/multiplication operators. Note that we use B to denote a component of K and not to denote the magnetic field from now on. For a series of equilibria, the operator C and A are positive, and the stability of the operator K is equivalent to the positivity of the following bilinear form $\mathcal{E}$:

$$\mathcal{E}\left(\begin{pmatrix} f \\ g \end{pmatrix}, \begin{pmatrix} f \\ g \end{pmatrix}\right) \equiv (Af, f) + (B^* f, g) + (Bg, f) + (Cg, g). \tag{2.2}$$

We remark here that we must be more precise to define $\mathcal{E}$ in each concrete cases. The above energy form $\mathcal{E}$ can be rewritten at least formally as:

$$\mathcal{E}(\begin{pmatrix} f \\ g \end{pmatrix}, \begin{pmatrix} f \\ g \end{pmatrix}) = (\{A - BC^{-1}B^*\}f, f) + (C\{g + C^{-1}B^*f\}, g + C^{-1}B^*f). \quad (2.3)$$

Using this expression, we can reduce the problem to the analysis of a single operator $D \equiv A - BC^{-1}B^*$. More precisely, assuming one of the axisymmetric, cylindrical and slab symmetries for the equilibrium and the shape of a region, we can define the operator A precisely. Then, we define the operators C^{-1}, $C^{-1}B$, B^*C^{-1}, $B^*C^{-1}B$ and D. We shall treat the case with a cylindrical symmetry in more detail in the next section. In general, the operator D is a non-negative second order differential operator with a non-negative top coefficient function which degenerates at "rational" magnetic surfaces. Now, the operator K is non-negative if and only if the energy form $\mathcal{E}$ is non-negative, and in this case, putting $g = -C^{-1}B^*f$, we have

$$\mathcal{E}(\begin{pmatrix} f \\ -C^{-1}B^*f \end{pmatrix}, \begin{pmatrix} f \\ -C^{-1}B^*f \end{pmatrix}) = (Df, f).$$

Hence the operator D is non-negative. Conversely if the operator D is non-negative, from the expression (2.3), the energy form $\mathcal{E}$ is non-negative.

3. Application to cylindrical case

In the case of cylindrical symmetry, we consider the problem in a region $\Omega \equiv \{(r, \theta, z) : 0 < r \leq r_0, 0 < \theta, z \leq 2\pi\}$. The concrete expression of the operators A,B,B^* and C is given as follows. The (m,k) Fourier modes are defined as a set of functions with the form:$e^{im\theta+ikz}v(r)$, where $v \in L^2(0, r_0; rdr)$. The force operator K in this (m,k) Fourier mode space is unitarily equivalent to the operator $K(k,m)$ in the function space $L^2(-\infty, s_0; ds)$ (see Kako[2]):

$$K(k,m) = \begin{pmatrix} A & B \\ B^* & C \end{pmatrix}$$

with

$$A = -\frac{d}{ds}\frac{b^2 + \gamma P}{r^2}\frac{d}{ds} + \left(b^2\frac{\sin^2\phi}{r^2}\right)' + b^2k_\phi^2, \quad (3.1)$$

$$B = \left(-i\frac{d}{ds}\frac{(b^2 + \gamma P)m_\phi}{r} + 2ib^2k\frac{\sin\phi}{r}, -i\frac{d}{ds}\frac{\gamma Pk_\phi}{r}\right), \quad (3.2)$$

$$B^* = \begin{pmatrix} -i\frac{(b^2+\gamma P)m_\phi}{r}\frac{d}{ds} - 2ib^2k\frac{\sin\phi}{r} \\ -i\frac{\gamma Pk_\phi}{r}\frac{d}{ds} \end{pmatrix}, \quad (3.3)$$

$$C = \begin{pmatrix} m_\phi^2(b^2 + \gamma P) + b^2k_\phi^2 & m_\phi k_\phi \gamma P \\ m_\phi k_\phi \gamma P & k_\phi^2 \gamma P \end{pmatrix}, \quad (3.4)$$

where $' = d/ds$ and

$$k_\phi = k(\cos\phi) + \frac{m}{r}(\sin\phi), \quad m_\phi = \frac{m}{r}(\cos\phi) - k(\sin\phi). \tag{3.5}$$

The coefficient functions b, ϕ and P are equilibrium quantities given as smooth functions which satisfy:

$$(P + \frac{b^2}{2})' + b^2\frac{\sin^2\phi}{r} = 0, \; ' = \partial/\partial r. \tag{3.6}$$

The operator $D = A - BC^{-1}B^*$ has the following form:

$$D = -\frac{d}{ds}f(s)\frac{d}{ds} + g(s) \tag{3.7}$$

with

$$f(s) = \frac{b^2k_\phi^2}{r^2k^2 + m^2} \tag{3.8}$$

and

$$g(s) = b^2k_\phi^2 + \left(\frac{b^2\sin\phi}{r^2}\right)' - \frac{4k^2b^2\sin^2\phi}{r^2k^2+m^2} + \left(\frac{2kb^2m_\phi\sin\phi}{r^2k^2+m^2}\right)', \tag{3.9}$$

where $r = e^s$. The function $f(s)$ vanishes quadratically at a rational surface where $k_\phi(s_c) = 0$. We call this s_c a rational point. We define a non-negative form $\mathcal{D}$ which corresponds to the non-negative operator D. Then the energy form $\mathcal{E}$ is defined as

$$\mathcal{E}(\begin{pmatrix} f \\ g \end{pmatrix}, \begin{pmatrix} \phi \\ \zeta \end{pmatrix}) \equiv \mathcal{D}(f,\phi) + (C^{1/2}\{g + C^{-1}B^*f\}, C^{1/2}\{\zeta + C^{-1}B^*\phi\}) \tag{3.10}$$

in some appropriate form domain (see Kako [2] for the exact treatment), and we can justify the rather formal procedure explained in the last section.

Further, we can deduce the so-called Suydam criterion as a corollary of this reduced analysis. Namely, using the fact that the rational point s_c is a regular singular point for the differential operator D, we can show that the operator D has an essential spectrum $[\{\frac{1}{4} + \frac{g(s_c)}{f''(s_c)/2}\}f''(s_c)/2, \infty)$. Hence the operator D is not non-negatice if the condition

$$\mu \equiv \frac{1}{4} + \frac{g(s_c)}{f''(s_c)/2} < 0, \tag{3.11}$$

is satisfied at the rational point s_c. This means that the following condition which is called the Suydam condition (see [4]):

$$\mu = \frac{1}{4} + \frac{g(s_c)}{f''(s_c)/2} \geq 0 \tag{3.12}$$

is necessary for the linearized stability. We call the value μ in (3.11) and (3.12) the Suydam index.

Lastly we shall give some brief comments on the difficulty in testing numerically the non-negativity of the operator D. For the general reference on the spectral approximation for the linearized MHD problem, see, for example, Gruber-Rappaz [1] and Kako-Descloux [3]. If we calculate approximately the spectrum by means of the FEM (Finite Element Method) using a space of piecewise linear continuous basis functions, we have the error estimate betweeen the lowest eigenvalue λ_0 of the operator D and its approximation $\lambda_{0,h}$:

$$|\lambda_0 - \lambda_{0,h}| \leq Ch^{\min\{2,2\sqrt{\eta}\}}, \tag{3.13}$$

in case $\mu = 0$, where a parameter η in the convergence order is defined as an absolute value of the possible negative lowest eigenvalue of the operator D. Hence the convergence of the approximate eigenvalues could be very slow when the Suydam index is zero and the system is slightly unstable which is the most important case in the optimization problem for the MHD system.

References

1. R. Gruber and J. Rappaz, "Finite element methods in linear magnetohydrodynamics," Springer-Verlag, Berlin Heidelberg, 1985.
2. T.Kako, *Essential spectrum of linearized operator for MHD plasma in cylindrical region*, J. Appl. Math. Phys. (ZAMP) 38 (1987), 433–449.
3. T Kako and J. Desclous, *Spectral approximation for the linearized MHD operator in cylindrical region*, Japan J. Indust. Appl. Math. 8 (1991), 221–244.
4. A. Lifschitz, "Magnetohydrodynamics and spectral theory," Kluwer Academic Publ., Dordrecht, 1989.

Discontinuity-Adaptive Godunov Scheme

C.Y. LEPAGE
Department of Applied Mathematics
University of Waterloo
Waterloo, Ontario
Canada
N2L 3G1

and

W.H. HUI
Department of Mathematics
The Hong Kong University of Science & Technology
Clear Water Bay, Kowloon
Hong Kong
Fax: 852-358-1643

Abstract

The celebrated Godunov scheme has been a foundation for shock-capturing computational methods for non-linear hyperbolic conservation laws, in particular the Euler equations of gas dynamics. Its main source of error, which arises from the averaging of the flow quantities over a computational cell, can be large if the cell contains discontinuities, namely shocks and sliplines. Most existing shock-capturing schemes for steady two-dimensional supersonic flow are based on the Eulerian formulation and tend to smear discontinuities, sliplines in particular, over several cells.

This paper introduces a discontinuity-adaptive Godunov scheme based on the Generalized Lagrangian Formulation. The crisp resolution of sliplines is inherent to the Generalized Lagrangian Formulation which uses streamlines and distance lines as coordinate lines. The sharp resolution of shocks, on the other hand, is achieved by imposing – through the splitting of shock-cells – the exact solution to the local Riemann problem, thus avoiding averaging over shocks.

Comparisons show that the discontinuity-adaptive Godunov scheme is superior to the classical one in both accuracy and efficiency.

1 The Generalized Lagrangian Formulation

Recently, Hui and his collaborators [1, 2, 3, 4] have introduced a Generalized Lagrangian Formulation (GLF) for the Euler equations of gas dynamics. The governing equations for steady two-dimensional flow of a perfect gas are obtained from their counterparts in the Eulerian formulation by transforming the Cartesian coordinates (x, y) to the Lagrangian-like coordinates (λ, ξ) by means of the transformation

$$dx = \frac{u}{q}d\lambda + U d\xi \qquad dy = \frac{v}{q}d\lambda + V d\xi \tag{1}$$

where $q = \sqrt{u^2 + v^2}$ is the speed of the flow, and U and V have dimensions of the reciprocal of the velocity. Physically, ξ represents a stream function while λ is the distance along a streamline. The transformed equations are

$$\frac{\partial \mathbf{E}}{\partial \lambda} + \frac{\partial \mathbf{F}}{\partial \xi} = 0 \tag{2}$$

where

$$\mathbf{E} = \begin{pmatrix} K \\ KH \\ Ku + pV \\ Kv - pU \\ U \\ V \end{pmatrix}, \qquad \mathbf{F} = \begin{pmatrix} 0 \\ 0 \\ -p\sin\theta \\ p\cos\theta \\ -\cos\theta \\ -\sin\theta \end{pmatrix},$$

$$K = \rho(uV - vU),$$

$$H = \frac{1}{2}(u^2 + v^2) + \frac{\gamma}{\gamma - 1}\frac{p}{\rho},$$

$$\sin\theta = \frac{u}{q}, \quad \text{and} \quad \cos\theta = \frac{v}{q},$$

θ being the flow inclination angle.

The use of streamlines as coordinate lines eliminates the need for grid generation since a solid boundary, itself a streamline of the flow, can be represented as a fixed coordinate line. Also, the coordinate λ gives a natural parametrization of a boundary in terms of its arclength.

For supersonic flow, the governing equations (2) are hyperbolic [4], and the classical Godunov scheme for marching in the λ direction is applicable. Its only source of error arises from averaging the flow quantities over a computational cell when representing the flow in all cells by piecewise constant states. In the smooth flow region the error is not large, being $\mathcal{O}(h)$ where h is the cell width, and it can be further reduced by applying the well-known TVD or ENO schemes. However, the error can be large and may lead to the non-existence of the solution to the Riemann problem if the cell contains discontinuities, namely shocks and sliplines.

2 Treatment of Discontinuities

The use of streamlines as coordinate lines in the Generalized Lagrangian formulation makes it possible to resolve sliplines crisply since a slipline, itself a streamline, can never cut across a cell. This is not the case in the Eulerian formulation where the slipline generally cuts across cells and the Godunov averaging then smears the discontinuity over several cells. In most cases, sliplines are known **a priori** and can be represented by fixed coordinate lines along the initial distance line. If this is not the case, the slipline will be confined to lie within a cell and its numerical resolution will still be very crisp.

To avoid the large error due to averaging the flow variables across a shock, the shock-cell, a computational cell containing both the upstream and downstream parts of a shock, is literally split into two subcells such that one of them is entirely upstream of the shock while the other one lies entirely downstream of the shock. The interface of the two subcells becomes a new coordinate line, a partition, which is used to track the evolution of the shock. The states of the two subcells are initially obtained from the solution of the local Riemann problem. The cell-splitting concept is easily generalized to model complex shock interactions since any subcell can be recursively split by inserting a new partition which moves at the speed of the new shock.

Only strong shocks are represented by partitions and tracked. A strong shock can be detected by looking, for example, at the pressure jump across an elementary shock wave identified in the solution to the local Riemann problem. The strong shock threshold may vary with the implementation.

Marching in the λ direction is as usual, except for minor considerations outlined below. In addition to satisfying the CFL linear stability condition, the step size $\Delta\lambda^n$ is adjusted such that the intersections of coordinate lines are hit exactly by distance lines. For example, the intersection of a shock with another shock or slipline can be modeled exactly by relaying all the programming decisions to the Riemann solver. Once the global step size has been determined, the conserved flow states $\mathbf{E}_j^n$ of all cells are advanced using the updating formula

$$\mathbf{E}_j^{n+1} = \mathbf{E}_j^n - \frac{\Delta\lambda^n}{\Delta\xi_j^{n+1}} \left\{ \mathcal{F}_{j+1/2}^{n+1/2} - \mathcal{F}_{j-1/2}^{n+1/2} \right\} \tag{3}$$

where

$$\mathcal{F}_{j+/1/2}^{n+1/2} = \mathbf{F}_{j+1/2}^{n+1/2} + \omega_{j+1/2} \left(\mathbf{E}_j^n - \mathbf{E}_{j+1/2}^{n+1/2} \right) \tag{4}$$

is the composite flux, $\omega_{j+1/2}$ is the slope $\frac{d\xi}{d\lambda}$, or speed, of the cell boundary $j+1/2$, $\mathbf{F}_{j+1/2}^{n+1/2}$ is the Godunov flux obtained from solving the local Riemann problem at $\xi_{j+1/2}$, and $\mathbf{E}_{j+1/2}^{n+1/2}$ is the conserved flow state along the inner side of the interface $j+1/2$, relative to the updated cell. This extension of the Godunov updating formula is also valid for the trapezoidal-shaped shock-cells, with special attention to the cell width which is the width of the cell at the new distance level λ^{n+1}. The composite flux reduces to the classical Godunov flux for regular cells for which $\omega = 0$.

The speed of each partition is also updated at each step by the speed of the corresponding elementary wave in the solution of the Riemann problem at the partition. Partitions are eliminated as they merge with fixed coordinate lines or with other partitions, the latter case being of special interest as it can only result from the intersection of two or more shocks. As is well known, a physical slipline will emerge from the intersection point of the two shocks. This new slipline can be captured exactly by introducing a new coordinate streamline as indicated by the merger of the two partitions.

3 Test Problem

To illustrate the effectiveness of the discontinuity-adaptive Godunov scheme in resolving shocks and sliplines crisply, the intersection of two shocks, whose flow deflection angles are 10° and −20°, respectively, is simulated. Comparisons of computed density profiles downstream of the intersection point of the two shocks are presented using the classical Godunov scheme based on the Eulerian formulation and the discontinuity-adaptive Godunov scheme based on the Generalized Lagrangian formulation. (See Figure 1.)

4 Conclusions

The shock treatment in the adaptive Godunov scheme differs from a shock-fitting scheme in that it is embedded in the Godunov scheme and it uses the solution of the local Riemann problems to detect shock discontinuities, without **a priori** knowledge of the existence nor position of the shock waves. Sliplines, being natural coordinate lines in the GLF, do not require any special treatment. In concordance with recent literature, this scheme should be referred to as a front-tracking method, or simply as a shock-tracking method as only shocks are actively tracked.

For simple flows, the discontinuity-adaptive Godunov scheme can reproduce the exact solution, as demonstrated by the above test problem. In other flows, the resolution of discontinuities is still very sharp. However, the scheme is still first order in regions of smooth flow.

Finally, the discontinuity-adaptive Godunov scheme is straightforward to program, using a linked list, by virtue of the simple structure of the solution of the Riemann problem in the (λ, ξ) coordinates. It was also found to be more efficient. For example, the average number of iterations per Riemann problem solved is down from about 1.5 to about 1.0 for the above problem.

Acknowledgements

This research was funded by grants from the U.S. NASA Lewis Research Center and the Natural Science and Engineering Research Council of Canada awarded to W.H. Hui. C.Y. Lepage was funded by an NSERC Graduate Scholarship.

References

[1] C.Y. Loh and W.H. Hui. "A New Lagrangian Method for Steady Supersonic Flow Computation. I. Godunov Scheme", *J. Comput. Phys.* **89**, 207 (1990).

[2] W.H. Hui and C.Y. Loh. "A New Lagrangian Method for Steady Supersonic Flow Computation. II. Slipline Resolution", *J. Comput. Phys.* **103**, 450 (1992).

[3] W.H. Hui and C.Y. Loh. "A New Lagrangian Method for Steady Supersonic Flow Computation. III. Strong Shocks", *J. Comput. Phys.* **103**, 465 (1992).

[4] W.H. Hui and Y.C. Zhao. "A Generalized Lagrangian Method for Solving the Euler Equations", Proceedings of the 4th International Conference on Hyperbolic Problems (to appear in *Notes on Numerical Fluid Mechanics*, by Vieweg).

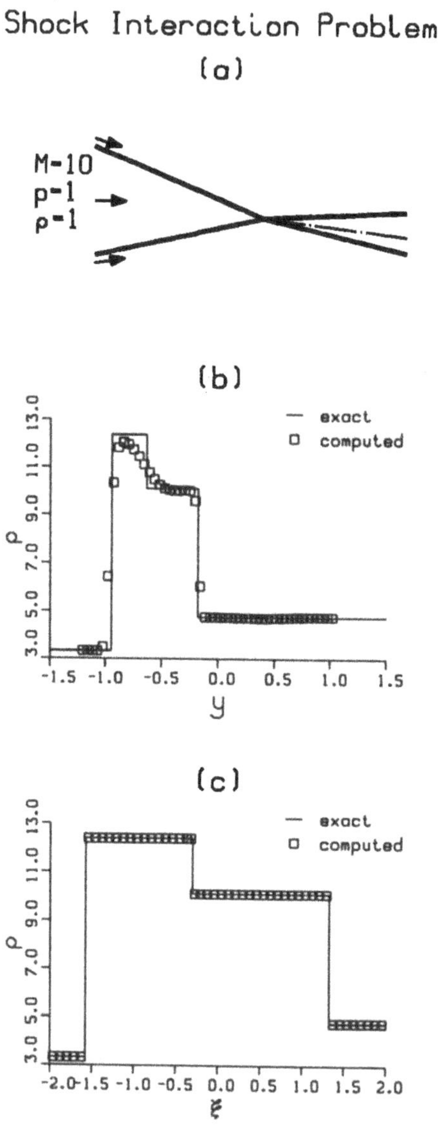

Figure 1: Density distribution in the shock interaction problem. (a) sketch, (b) Godunov scheme based on Eulerian formulation, (c) Adaptive Godunov scheme based on the Generalized Lagrangian Formulation

Automatically Capturing Technique of Discontinuous Solution of Euler Equations

Li Song-bo Wang Yun-tao
China Aerodynamic Research and Development Center
Mian-yang (621000), Si-chuan, China

Abstract

A theoretical and numerical study was made on the essential reasons for abnormal phenomena, such as oscillations across shock wave, nonlinear instability and convergence to nonphysical solution, encountered in computational aerodynamics. A class of the least dissipation-correction difference schemes (LDC) of high accuracy were presented to overcome these difficulities. With four new standard one dimensional Riemann problems instead of the Sod's example, comparative numerical simulations were made by using the TVD schemes and the LDC schemes, which illustrate that among these schemes the LDC schemes are in the best agreements with the exact solutions.

Part I. Reasons for the three abnormal phenomena

The initial value problem of one dimensional Euler equations is expressed by following equations

$$U_t + F_x = 0 \tag{1.1}$$

$$U = (\rho, \rho u, E)^T, \quad F = (\rho u, \rho u^2 + p, uE + up)^T \tag{1.2}$$

$$U(x,0) = U_0(x) \tag{1.3}$$

In order to obtain the unique physical relevant solution, expect the Rankine-Hugoniot relations, Lax's entropy condition[6] must be considered as a supplementary condition.

1. Oscillations across shock wave. The following conclusion has been obtained in ref. [1–3]: The reason for shock oscillation is that near a shock wave, since the difference scheme is dissipation at one point, but negative dissipation at another which has unfixed location, a set of new local extreme generates as time going on, so that oscillations across the shock wave take place.

2. Nonlinear instability. The following conclusion has been got in ref. [1–3]: If the difference scheme is negative dissipation near sonic line or at stagnation point,

nonlinear "weak" instability takes place, which leads to small jump of ρ, p and u near these points. If the difference scheme is negative dissipation near a steady shock wave, nonlinear "strong" instability takes place, which make p_0^n quickly drop to negative one, p_1^n quickly rise to infinity.

3. Nonphysical phenomenon. we have got those conclusions in ref. [1–3]: First, the nonphysical solution obtained by the conservative difference schemes must contain a steady rarefaction shock wave (U_-, U_+) that satisfies the R-H relations, but does not satisfy Lax's entropy condition. Second, for the initial value of the steady rarefaction shock wave, the dissipative coefficients of the conservative schemes are equal to zero, then, they are zero dissipation and degenerate into identical operator, so that the numerical solutions are nonphysical. The last, nonphysical phenomena may only occur in the calculation of 1-rarefaction wave. For some TVD schemes, nonphysical phenomenon only appears in the calculation of 1-rarefaction wave containing the first generalized eigenvalues[5] with different signs.

Summarizing above conclusions,we found that it is the negative dissipation or zero dissipation of some difference schemes on some special points that cause these abnormal phenomena, therefore, the effective method of overcoming these difficulties is to use the dissipative difference scheme[1].

Part II. The least dissipation-correction schemes

By Roe's average[7], the least dissipation-correction schemes first proposed in ref. [4] can be rewritten in the form

$$U_j^{(1)} = U_j^n + \sum(a_{kj+1/2}^n d_{kj+1/2}^n r_{kj+1/2}^n - b_{kj-1/2}^n d_{kj-1/2}^n r_{kj-1/2}^n) \tag{2.1}$$

$$U_j^{n+1} = U_j^{(1)} - \sum(\mu_{kj+1/2} d_{kj+1/2}^{(1)} r_{kj+1/2}^{(1)} - \mu_{kj-1/2} d_{kj-1/2}^{(1)} r_{kj-1/2}^{(1)}) \tag{2.2}$$

where, $r_{kj+1/2}$ is the corresponding right eigenvector,

$$\begin{aligned} d_{1j+1/2} &= (\Delta p_j - c_{j+1/2}\sqrt{\rho_j \rho_{j+1}}\Delta u_j)/(2c_{j+1}^2) \\ d_{2j+1/2} &= \Delta p_j / c_{j+1/2}^2 - d_{1j+1/2} \\ d_{3j+1/2} &= \Delta \rho_j - d_{1j+1/2} - d_{3j+1/2} \end{aligned}$$

If (U_j, U_{j+1}) is a k-wave which satisfies the R-H relations, then

$$d_{kj+1/2} = \Delta\rho_j, \qquad d_{ij+1/2} = 0.0, \qquad i \neq k. \tag{2.3}$$

Various difference schemes are generated from Eq. (2.1)-(2.2) by making different choice of a_k, b_k, μ_k. Accordingly, we present four schemes here

(1) LDC 1. It is a symmetric LDC scheme

$$q_j = 1/2(|u_j| + c_j)\Delta t/\Delta x[1 - (|u_j| + c_j)\Delta t/\Delta x]$$

$$\begin{aligned}
q_{j+1/2} &= (q_j + q_{j+1})/2 \\
\mu_{0j+1/2} &= \sigma_{j+1/2}(1 - \theta_{j+1/2})q_{j+1/2} \\
\theta_{j+1/2} &= \text{MAX}\,(\theta_j, \theta_{j+1}, \Delta x) \\
\delta_j &= |\Delta\rho_{j-1}| + |\Delta\rho_j| \\
\theta_j &= \begin{cases} |\Delta|\Delta\rho_{j-1}||/\delta_j : & \delta_j \geq 0.05\rho_{\max} \\ 0.0 & \delta_j < 0.05\rho_{\max}; \end{cases}
\end{aligned}$$

where $\sigma_{j+1/2}$ is distinguishable operator defined as

$$\sigma_{j+1/2} = (|\Delta u_j^n| + |\Delta c_j^n|)/(|\Delta u_j^n| + |\Delta c_j^n| + 0.0001)$$

(2) LDC 2. This is a non-symmetric LDC scheme

$$\begin{aligned}
\alpha_{kj+1/2} &= \max(|\lambda_{kj+1/2}|\Delta t/\Delta x, 0.01) \\
a^n_{kj+1/2} &= 1/2(\alpha - \lambda\Delta t/\Delta x)^n_{kj+1/2} \\
b^n_{kj-1/2} &= 1/2(\alpha + \lambda\Delta t/\Delta x)^n_{kj-1/2} \\
\delta_{kj+1/2} &= |\Delta d_{kj-1/2}| + |\Delta d_{kj+1/2}| + 0.02 \\
q_{kj+1/2} &= 1/2|\lambda_{kj+1/2}|\Delta t/\Delta x(1.0 - |\lambda_{kj+1/2}|\Delta t/\Delta x) \\
\mu_{kj+1/2} &= \sigma_{j+1/2}(1.0 - |\Delta^2 d_{kj+1/2}|/\delta_{kj+1/2})q_{kj+1/2}
\end{aligned}$$

(3) LDC 3. This is a non-symmetric second-order LDC scheme, a, b, δ are the same as LDC 2 except $\mu_{kj+1/2}$ is defined as

$$\begin{aligned}
\mu_{1j+1/2} &= \sigma_{j+1/2}[q^n_{1j+1/2} - \min(q^n_{1j+1/2}, 4|\Delta\lambda^n_{1j+1/2}\Delta t/\Delta x|)] \\
\mu_{2j+1/2} &= \sigma_{j+1/2}[q^n_{2j+1/2} - \min(1/3 * q^n_{2j+1/2}, 6|\Delta c_j \Delta t/\Delta x|)] \\
\mu_{3j+1/2} &= \sigma_{j+1/2}[q^n_{3j+1/2} - \min(1/4 * q^n_{3j+1/2}, 8|\Delta\lambda^n_{3j+1/2}\Delta t/\Delta x|)]
\end{aligned}$$

(4) LDC 4. This is a symmetric second-order LDC scheme, q_j, $q_{j+1/2}$, $\delta_{j+1/2}$ are the same as LDC 1 except $\mu_{0j+1/2}$ is defined as

$$\mu_{0j+1/2} = \sigma_{j+1/2}[q_{j+1/2} - \min(\frac{1}{8}, (|\Delta u_j| + |\Delta c_j|)\Delta t/\Delta x)]$$

Part III. Standard examples and numerical results

The 1-D shock tube problem of Sod is usually adopted as standard example to examine difference schemes. According to our theory, both of its shock and rarefaction waves are too weak to be used as standard example. Thus, four new standard examples are presented in this paper for examination of difference schemes

Example 1. $p_- = 29, \quad \rho_- = 5, \quad u_- = \sqrt{1.4},$
$p_+ = 1.0, \quad \rho_+ = 1., \quad u_+ = 5\sqrt{1.4},$

Example 2. $p_- = 18.5, \quad \rho_- = 21.707, \quad u_- = 0.218,$
$p_+ = 1.0, \quad \rho_+ = 1., \quad u_+ = 4.732,$

Example 3. $p_- = 1.0, \quad \rho_- = 1.0, \quad u_- = 10\sqrt{1.4} + 4.0,$
$p_+ = 50, \quad \rho_+ = 36100/5234, \quad u_+ = 4.0,$

Example 4. $p_- = 1.0, \quad \rho_- = 1.0, \quad u_- = 10\sqrt{1.4},$
$p_+ = 100\sqrt{1.4}, \quad \rho_+ = 10\sqrt{1.4}, \quad u_+ = 5\sqrt{1.4}$

We strictly examined the first order upwind scheme[8] (Upwind1), TVD2[8], UNO2[8], NND1[9] and LDC 1 to LDC 4 with these examples. The numerical results of example 2 to example 4 are shown in Figure 1 to Figure 6, where the solid line represent the theoretical solution. Since the corresponding results of LDC 1 and LDC 2 are similar to that of LDC 4, they are not presented here.

For example 1, by substituting (U_-, U_+) into these schemes directly rather than doing numerical simulations, it is immediately found that the results obtained by Upwind1, TVD2, UNO2 are all nonphysical solutions. As to example 2, there is a steady rarefaction shock involved in the 1-simple fan calculating with Upwind1, TVD2. For example 3, TVD2, UNO2 and NND1 are all fail in calculation because of undershoot, only LDC scheme and Upwind1 are successful. For example 4, $p < 0.0$ take place calculating with NND1.

It can be observed from above results that these standard examples are rigorously and scientific, which can examine the reliability and applicable range of difference schemes. For these standard example, only LDC schemes are universal, with which high accuracy is obtained for the strong rarefaction wave. Near zero point of the generalized eigenvalue, schemes of other types may give either a steady rarefaction shock wave, or a small jumps of ρ, p and u, besides the poor accuracy near the endpoints of rarefaction wave. This verifies, on the other hand, our conclusions that the essential reason for these abnormal phenomena is that the difference scheme is negative dissipation or zero dissipation on some special points.

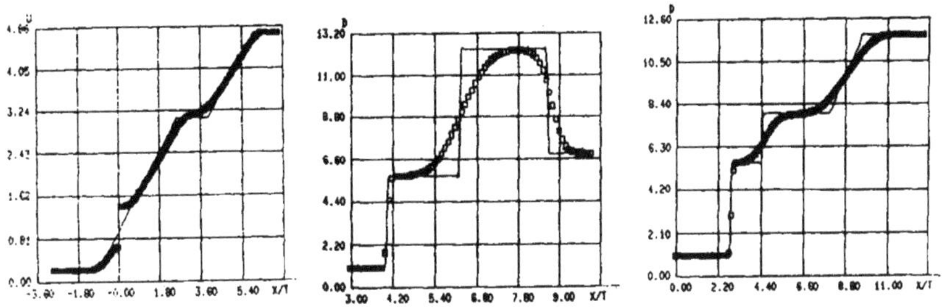

Fig. 1 Numerical results of example 2-4 with Up-wind1.

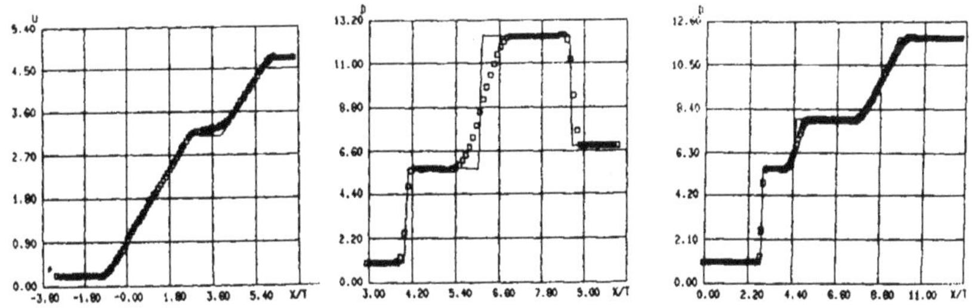

Fig. 2 Numerical results of example 2-4 with LDC 3

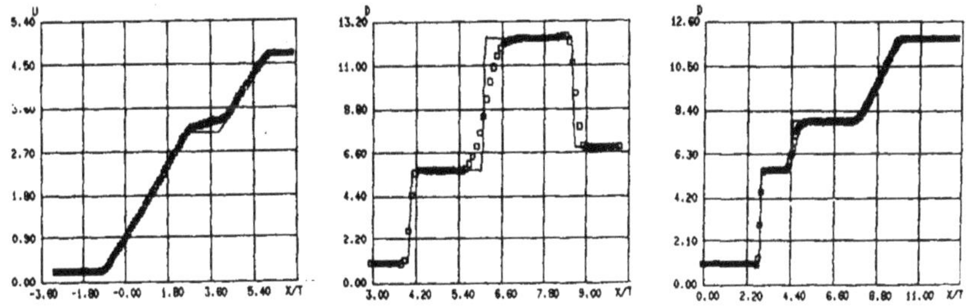

Fig. 3 Numerical results of example 2-4 with LDC 4

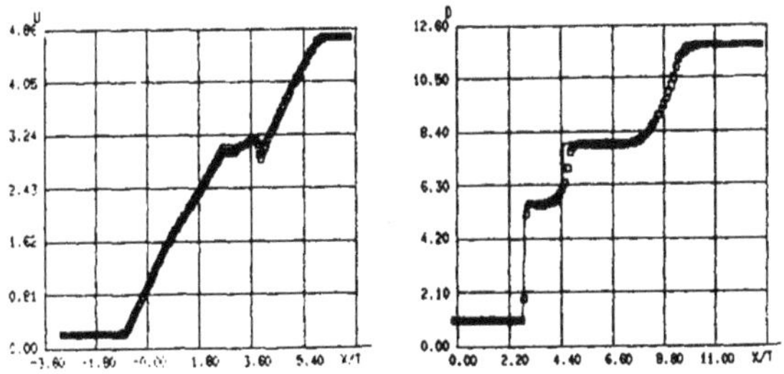

Fig. 4 Numerical results of example 2 and exmple 4 with UNO 2

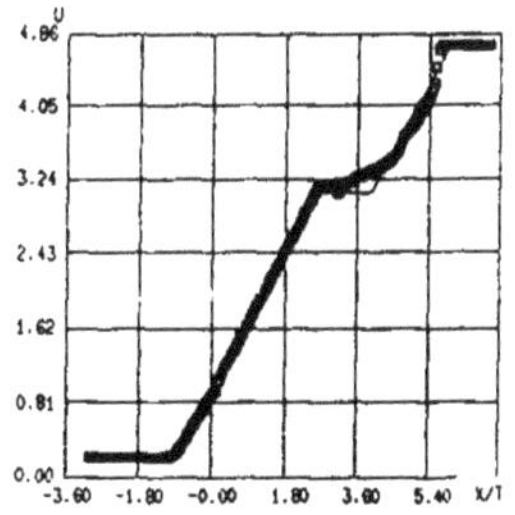

Fig. 5 Example 2 with NND 1

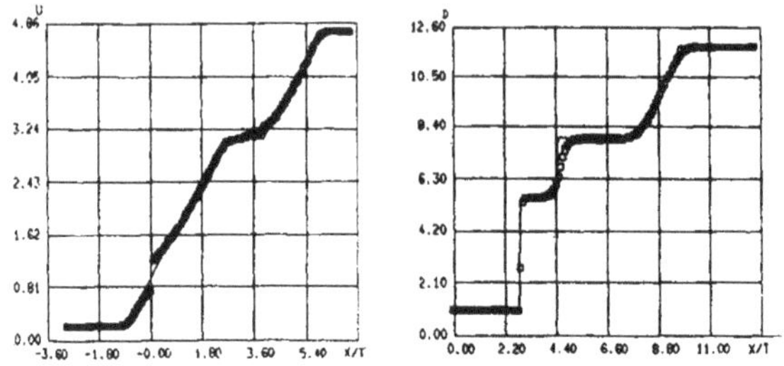

Fig. 6 Numerical results of example 2 and exmple 4 with TVD 2

References

1. Li Song-bo, *The proceedings of the computational mathematics conference*, Tianjin, 1991 (in press).
2. Li Song-bo, *The proceedings of the 5th conference on numerical methods for computational fluid dynamics*, Mianyang, 1991 (in press).
3. Li Song-bo, *Acta Aerodynamica Sinica*, Vol. 1, No. 4, 1983, pp. 1–9.
4. Li Song-bo, *J. Comp. Math*, Vol. 9, No. 4, 1991, pp. 314–320.
5. Li Song-bo, *The proceedings of the 6th computational dynamics conference*, China, 1992.
6. P.D. Lax, *Comm. Pure Appl. Math.*, Vol. 10, 1957.
7. P.L. Roe, *J. Comp. Phys.*, Vol. 43, 1981, pp. 357–372.
8. A. Harten and S. Osher, *SIAM J. Num. Anal.*, Vol. 24, 1987, pp. 279–309.
9. Zhang Han-xin, *Acta Aerodynamica Sinica*, Vol. 6, No. 2, 1988, pp. 145–155.

SHOCK CAPTURING WITH MODIFIED COMPACT DIFFERENCE SCHEMES

Ma Yanwen, Fu Dexun

LNM, Institute of Mechanics, Chinese Academy of Sciences,Beijing, 100080, P. R.China

ABSTRACT: When the compact scheme is used to capture the shock there exist oscillations in the numerical solutions. The commpact scheme is modified with the method of diffusion analogy developed by authors. The developed method is used to compute the 2–D shock reflection,and good solutions are obtained.

1. Introduction

Great successes have been reached for the shock capturing, and different TVD type schemes are developed. The TVD schemes are second order accurate and the accuracy will be reduced at the extremal points.For numerical simulation of complex flow field with multi–scale structure the second order accuracy of schemes is not enough, and development of higher order accurate schemes are needed. It is known that the compact schemes are efficient, and they have high accuracy of approximation with less stencil of mesh points,but there exist large oscillations in the numerical solutions when they are used to capture the shock. It is difficult to adopt the method of constructing TVD and ENO type schemes for reconstructing the compact schemes to improve the resolution of the shock. The purpose of the present paper is to develop a new class of modified compact schemes which can capture the shock well.

2. Compact Difference Approximation

Consider the following model equation

$$\frac{\partial u}{\partial t} + \frac{\partial f}{\partial x} = 0, \qquad f = cu, \tag{1}$$

The corresponding kth order accurete semi–discrete approximation is written as

$$\frac{\partial u_j}{\partial t} + \frac{F_j}{\Delta x} = 0 \tag{2}$$

where F_j is obtained from the following algebraic equation

$$\sum_m \alpha_m F_{j+m} = \sum_m \beta_m (f_{j+m+1} - f_{j+m}) \tag{3}$$

This work is supported by Key Project for Basic Researches

Denote

$$\delta_x^{\pm} f_j = \pm(f_{j\pm1} - f_j), \quad \delta_x^0 = \frac{1}{2}(\delta_x^+ + \delta_x^-), \quad \delta_x^2 = \delta_x^+ \delta_x^- \tag{4}$$

Several differences are listed below:

Second order accurate difference: $F_j = \delta_x^0 f_j$

Fourth order accurate difference: $F_j = \delta_x^0 f_j - \frac{1}{12}(\delta_x^2 f_{j+1} - \delta_x^2 f_{j-1})$

Fourth order accurate compact difference: $\frac{1}{6}F_{j+1} + \frac{2}{3}F_j + \frac{1}{6}F_{j-1} = \delta_x^0 f_j$

Sixth order accurate compact difference:

$$\frac{1}{5}F_{j+1} + \frac{3}{5}F_j + \frac{1}{5}F_{j-1} = \frac{14}{15}\delta_x^0 F_j + \frac{1}{30}(\delta_x^0 f_{j+1} + \delta_x^0 f_{j-1})$$

It was shown in ref.[3] that the higher order accurate difference approximation can approximate the derivatives of Fourier components with higher wave number more correctly. This is very important for correctly simulating the multi–scale structure of flow field. It also can be shown that with the same order of accuracy of approximation the compact difference can give better approximation with wider band of wave numbers.

3. Method of Diffusion Analogy for Constructing High Order Accurate Schemes

The solution of the scale conservation law has TVD property. The TVD property is not common for the solutions of the nonlinear hyperbolic system of equations. It can be shown that some kind of artificial viscosity is introduced for the TVD type schemes. It means that the reason of improvement of the shock solution for the TVD type schemes is implicitly introduced viscosity. The quastion is what kind of artificial viscosity need to be introduced. Generally speaking, the type of the introduced artificial viscosity depends on the scheme, the order of accuracy of approximation and the behaviour of the solutions.

In ref. [1,2] a second order accurate scheme obtained with the method of diffusion analogy is used to solve the 2-D shock reflection problem. This method of diffusion analogy is used to reconstruct the compact schemes in the present paper to obtain high resolution of the shock.

According to the modified equation for the k th order accurate approximation (2) we can obtain the following diffusion analogy equation

$$\frac{\partial u}{\partial t} + \frac{\partial f}{\partial x} = \Delta x \frac{\partial}{\partial x}\tilde{\mu}\frac{\partial u}{\partial x} \tag{5}$$

where $\tilde{\mu}$ is called a diffusion analogy coefficient and is defined as

$$\tilde{\mu} = \sigma \triangle x^{k-1}\frac{\partial^k u}{\partial x^k}/\frac{\partial u}{\partial x}c. \tag{6}$$

where σ is a parameter which depends on the used scheme. The oscillations may be produced in the numerical solutions in region where $\tilde{\mu}$ is negative.Suppose $\sigma c > 0$. From expression (6) it can be seen that the sign of $\tilde{\mu}$ depends on $\frac{\partial^k u}{\partial x^k}/\frac{\partial u}{\partial x}$. Suppose we have a N-S shock which has a steep gradient in the vicinity of S(Fig.1) and necessary derivatives. It's derivatives are given in Fig.2. From Fig.2 we see that for the case $k = 1, \tilde{\mu} = \sigma c > 0$, this corresponds to the 2-point upwind difference. In the case k=2 and $c\sigma > 0$, behind the shock $\tilde{\mu} > 0$ and in front of the shock $\tilde{\mu} < 0$. It means that oscillations may be produced in front of the shock. In the case $k \geq 3$ in both sides of the shock exist subregions where $\tilde{\mu} < 0$, so oscillations may be produced in both sides of the shock.

Denote a positive $\bar{\mu}$ as a reconstructed coefficient, the modified scheme is defined as

$$\frac{\partial u_j}{\partial t} + \frac{F_j}{\Delta x} = \frac{1}{\Delta x}\left[\bar{\mu}_{j+\frac{1}{2}}\delta_x^+ u_j - \bar{\mu}_{j-\frac{1}{2}}\delta_x^- u_j\right] \tag{7}$$

4. Modified Compact Difference Approximation for the Euler Equations

Consider the 1-D Euler equations

$$\frac{\partial U}{\partial t} + \frac{\partial f}{\partial x} = 0 \tag{8}$$

$$U = [\varrho, \varrho u, E]^T \quad , \quad f = [\varrho u, \varrho u^2 + p, u(E+p)]^T$$

$$p = \frac{1}{\gamma M_\infty^2}\varrho T \quad , \quad E = \varrho(C_v T + \frac{u^2}{2}) \tag{9}$$

The lst derivative of the flux vector is approximated with the fourth order accurate compact difference. The corresponding diffusion analogy equations and the coefficient are defined as

$$\frac{\partial U}{\partial t} + \frac{\partial f}{\partial x} = \Delta x \frac{\partial}{\partial x}\tilde{M}\frac{\partial U}{\partial x} \tag{10}$$

$$\tilde{M} = \sigma \tilde{D} A \tag{11}$$

where A is the Jacobian matrix, $\tilde{D}$ is a diagonal matrix with elements

$$\tilde{D}_{i,i} = \Delta x^3 \frac{\partial^4 f_i}{\partial x^4} / \frac{\partial f_i}{\partial x} \tag{12}$$

f_i is the i-th component of the flux vector f. The reconstructed scheme is defined as

$$\frac{\partial U_j}{\partial t} + \frac{F_j}{\Delta x} = \frac{1}{\Delta x}\left[\bar{M}_{j+\frac{1}{2}}\delta_x^+ U_j - \bar{M}_{j-\frac{1}{2}}\delta_x^- U_j\right] \tag{13}$$

The modified matrix coefficient $\bar{M}$ is defined as

$$\bar{M} = |\sigma|\bar{D}|A| \quad , \qquad |A| = S^{-1}|\wedge|S \tag{14}$$

S^{-1} and S are the matrices of similarity transformation, $\wedge$ is the diagonal matrix with

$$\lambda_1 = u, \quad \lambda_2 = u - c, \quad \lambda_3 = u + c \tag{15}$$

where c is the sound speed, $|\wedge|$ is the diagonal matrix with $|\lambda_m|$. The diagonal matrix $\bar{D}$ is obtained from $\tilde{D}$ after reconstruction. Expression (12) can be rewritten as

$$\tilde{D}_{i,j} = \Delta x^3 \left(\frac{\partial^2 f_i}{\partial x^2} / \frac{\partial f_i}{\partial x}\right)\left(\frac{\partial^3 f_i}{\partial x^3} / \frac{\partial^2 f_i}{\partial x^2}\right)\left(\frac{\partial^4 f_i}{\partial x^4} / \frac{\partial^3 f_i}{\partial x^3}\right) \tag{16}$$

$$\tilde{D}_{i,j} = \Delta x^3 \frac{\partial}{\partial x} Ln\Big(\frac{\partial f_i}{\partial x}\Big) \cdot \frac{\partial}{\partial x} Ln\Big(\frac{\partial^2 f_i}{\partial x^2}\Big) \cdot \frac{\partial}{\partial x} Ln\Big(\frac{\partial^3 f_i}{\partial x^3}\Big) \tag{17}$$

Three methods of reconstruction of the matrix $\tilde{D}$ are presented:

1) Taking absolute value

Expression (12) can be approximated at the point $j + \frac{1}{2}$, for example, as follows

$$(\bar{D}_{i,i})_{j+\frac{1}{2}} = \frac{3|\delta_x^4 (f_i)_{j+\frac{1}{2}}|}{|\delta_x^+ (f_i)_{j+1}| + |\delta_x^+ (f_i)_j| + |\delta_x^+ (f_i)_{j-1}|} \tag{18}$$

where δ_x^4 is a fourth difference

2) Using simple minmod function

The right hand side of (16) can be approximated as

$$(\bar{D}_{i,i})_{j+\frac{1}{2}} = \Big[1 - (\gamma_i^{1,-})_{j+\frac{1}{2}}\Big]\Big[1 - (\gamma_i^{2,-})_{j+\frac{1}{2}}\Big]\Big[1 - (\gamma_i^{3,-})_{j+\frac{1}{2}}\Big] \tag{19}$$

where

$$(\gamma_i^{1,\pm})_{j+\frac{1}{2}} = \frac{\delta_x^+ (f_i)_{j\pm 1}}{\delta_x^+ (f_i)_j}, \quad (\gamma_i^{2,\pm})_{j+\frac{1}{2}} = \Big[\frac{\delta_x^2 (f_i)_{j+1}}{\delta_x^2 (f_i)_j}\Big]^{\pm 1}, \quad (\gamma_i^{3,\pm})_{j+\frac{1}{2}} = \frac{\delta_x^3 (f_i)_{j+\frac{1}{2}\pm 1}}{\delta_x^3 (f_i)_{j+\frac{1}{2}}}$$

For making (19) positive the minmod function can be used, for example

$$(\bar{D}_{i,i})_{j+\frac{1}{2}} = \Big[1 - (Q_i^1)_{j+\frac{1}{2}}\Big]\Big[1 - (Q_i^2)_{j+\frac{1}{2}}\Big]\Big[1 - (Q_i^3)_{j+\frac{1}{2}}\Big] \tag{20}$$

$$(Q_i^n)_{j+\frac{1}{2}} = \text{minmod}\Big\{1, (\gamma_i^{n,+})_{j+\frac{1}{2}}, (\gamma_i^{n,-})_{j+\frac{1}{2}}\Big\} \tag{21}$$

Other minmod function can be used.

3) Using logarithm-minmod function

Expression (17) can be approximated at the point $(j + \frac{1}{2})$ as

$$(\tilde{D}_{i,i})_{j+\frac{1}{2}} = Ln(\gamma_i^{1,+})_{j+\frac{1}{2}} \cdot Ln(\gamma_i^{2,+})_{j+\frac{1}{2}} \cdot Ln(\gamma_i^{3,+})_{j+\frac{1}{2}}$$

The modified component $\bar{D}_{i,j}$ can be defined as

$$(\bar{D}_{i,i})_{j+\frac{1}{2}} = -Ln(\bar{Q}_i^1)_{j+\frac{1}{2}} \cdot Ln(\bar{Q}_i^2)_{j+\frac{1}{2}} \cdot Ln(\bar{Q}_i^3)_{j+\frac{1}{2}} \tag{22}$$

$$(\bar{Q}_i^n)_{j+\frac{1}{2}} = \max\Big\{\varepsilon, (Q_i^n)_{j+\frac{1}{2}}\Big\}, \varepsilon = 0.3679 \tag{23}$$

Expression (21) can be used for defenition of $(Q_i^n)_{j+\frac{1}{2}}$. Introdnction of the parameter ε is to avoide singularity of the logarithm function.

If the elements of $|\wedge|$ are replaced by $\lambda = \max|\lambda_m|$, the matrix $\bar{M}$ becomes diagonal

$$\bar{M} = |\sigma|\bar{D}\lambda$$

and the computer work can be reduced considerably.

5. Numerical Examples

The above presented method is used to compute the shock reflection problem with $M_\infty = 2.9$ and incident shock angle $\theta = 29°$. The computed pressure distribution on $Y = 0.5$ and the pressure contours are given in Fig.3. Fig.3a is obtained with the first method of reconstruction of the diffusion analogy coefficient; Fig.3b shows the results with simple minmod function, the results in Fig.3c are obtained with logarithm-minmod function.

REFERENCES

1 Fu Dexun Ma Yanwen, 4th International Symposium on Computational Fluid Dynamics, Davis, 1991

2 Ma Yanwen, Fu Dexun, Science in China Series A, Vol. 35, No.9 1992

3 LeLe, S. K., AIAA 89-0374 1989

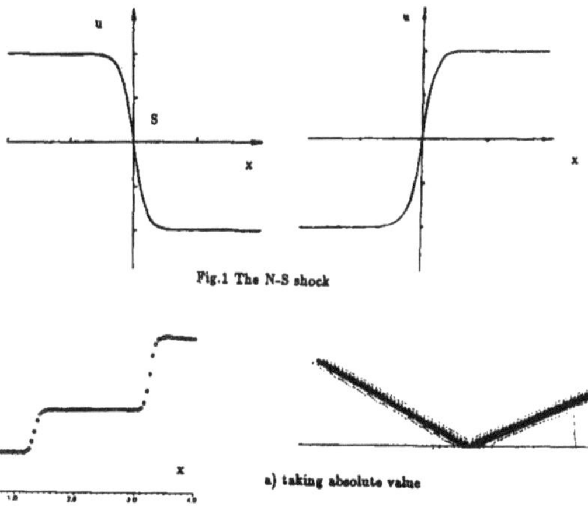

Fig.1 The N-S shock

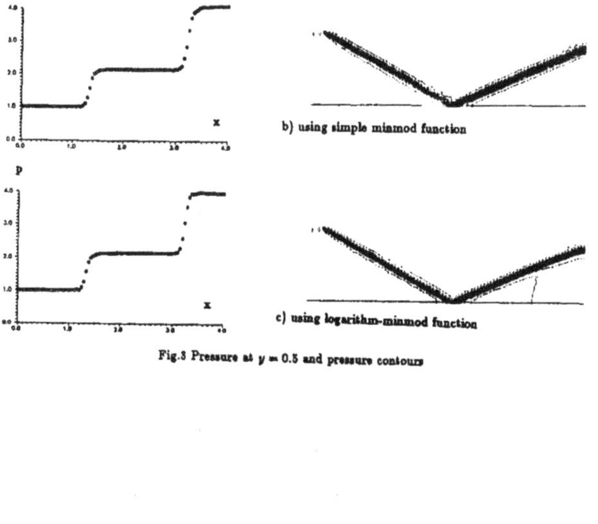

a) taking absolute value

b) using simple minmod function

c) using logarithm-minmod function

Fig.3 Pressure at $y = 0.5$ and pressure contours

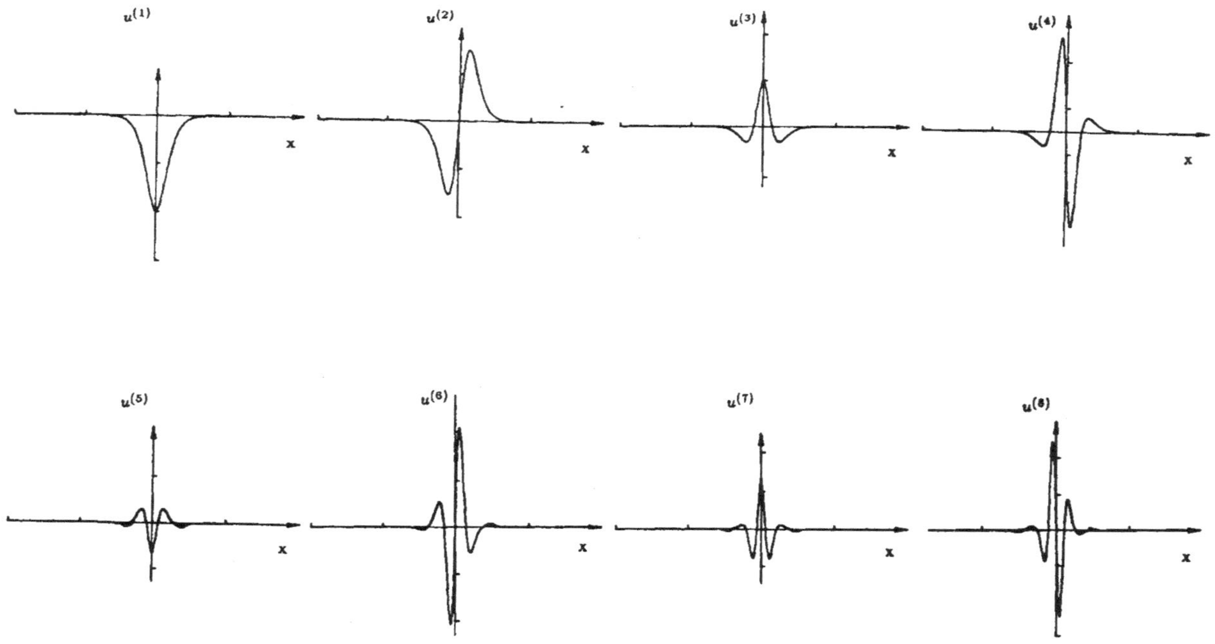

Fig.2 Derivatives of the N-S shock

Hierarchy of Uniformly High Accuracy and High Resolution Schemes for Hyperbolic Equations[1]

HUAMO WU
LSEC, COMPUTING CENTER, BEIJING 100080, CHINA

Abstract. In this paper we present a new class of high resolution schemes using minimal stencil and having maximal order of accuracy. The entropy conditins and the conservation laws are satisfied asymtotically when Δx and Δt tend to zero. Numerical results for model problems are given. Experiments show the uniformly high accuracy of expected order for smooth solutions and high resolution abilities for discontinuities

1. Introduction. In the last decades the technique of high resolution schemes such as Harten's total variation diminishing (TVD) schemes [1] , Shu's total variation bounded (TVB) schemes [2] or uniformly high accurate and non-oscillatory (UNO or ENO) schemes [3] becomes a powerful tool in CFD applications and a challenging subject of theoretical studies. So far, most of the proposed and used schemes for conservation laws are conservative which may guarantee the proper shock motion and ensure the numerical solutions to be weak solutions by using the celebrated Lax - Wendroff theorem [4] on weak solutions. It is well known that explicit conservative high resolution schemes such as TVD [1], ENO [2,3] etc. schemes of second order require 5 or 7 nodes in the stencil of the schemes. The schemes of accuracy order higher than 2 require even more stencil points. This fact makes the computation near boundaries difficult. On the other hand, the usual nonconservative schemes may require less points in the stencil, and hence resolve this problem easier,but may cause wrong shock locations and violate the entropy conditions due to the nonconservative nature.

In this paper, we give an approach of building a hierarchy of schemes which are uniformly high order with conservation properties and entropy conditions satisfied asymptotically when the meshsize tends to zero.

The basic idea to construct such kind of schemes is based on our earier work [5] and recent results [6] and [7]. Firstly, we should have a good first order conservative scheme

$$u^{n+1} = L(u^n) \tag{1.1}$$

[1]Project supported by The National Natural Science Foundation of China and The State Commission of Science and Technology of China

which is an E - scheme [8] in the case of scalar conservation laws and to be the schemes of Lax-Friedrichs, Godunov, or Rusanov etc. for systems to insure the entropy conditions to be satisfied. At the same time, we should have a second order accurate high resolution scheme

$$u^{n+1} = S(u^n) \tag{1.2}$$

which is either conservative [9] or nonconservative [10], and a very high accurate scheme

$$u^{n+1} = H(u^n). \tag{1.3}$$

Next, it is very important to develope a technique of choosing a scheme from (1.1),(1.2), and (1.3) to be used at each grid point.

In the next section we give the description of such a technique and examples of fourth order uniform accurate and high resolution schemes using 5 point stencil. The numerical results are given in the last section.

2. The Schemes. Let $\Delta_+ u, \Delta_- u$ be the backward and forward differences of u, Δx be the spatial stepsize. Let M, C, A, and α be possitive constants (M=5, C=1.8,A=1., α=0.8 for example). Introducing the quantities:

$$\begin{aligned}
&SL(u) = S(u) - L(u),\\
&HL(u) = H(u) - L(u),\\
&AD = max(1., \left|\frac{\Delta_+ u}{\Delta x}\right|^{C-1}, \left|\frac{\Delta_- u}{\Delta x}\right|^{C-1})\\
&Q = min(M * AD * \Delta x^C, SL) * sign(SL)\\
&SC = \begin{cases} HL, & \text{if } |HL| < A * \Delta x^\alpha \\ Q, & \text{otherwise} \end{cases}
\end{aligned} \tag{2.1}$$

we have the following scheme

$$u^{n+1} = L(u^n) + R(u^n). \tag{2.2}$$

The nature of technique (2.1) is to suppress the unwanted oscillations of the high order accurate schemes near steep gradients, keeping the high accuracy of the schemes in the region of smoothness of the solutions. As an example of our constructed schemes, we present an uniformly four order accurate scheme using 5 points in the stencil for the model equation

$$\frac{\partial u}{\partial t} + a\frac{\partial u}{\partial x} = 0, \quad a > 0 \tag{2.3}$$

Let $\nu = a\Delta t/\Delta x$ be the CFL number, $\Delta^k u$ be the k-th differences, namely,

$$\begin{aligned} &\Delta^1_{j-1/2}u \equiv \Delta_{j-1/2}u = u_j - u_{j-1}, \quad &\Delta^2_j u = \Delta_{j+1/2}u - \Delta_{j-1/2}u, \\ &\Delta^3_{j-1/2}u = \Delta^2_j u - \Delta^2_{j-1}u, \quad &\Delta^4_j u = \Delta^3_{j+1/2}u - \Delta^3_{j-1/2}u. \end{aligned} \tag{2.4}$$

Define four schemes P_1, P_2, P_3, P_4 of first to fourth order of accuracy respestively:

$$\begin{aligned} u_j^{n+1} &= P_1(u^n) = u_j^n - \nu\Delta_{j-1/2}u^n, \\ u_j^{n+1} &= P_2(u^n) = u_j^n - \nu\Delta_{j-1/2}u^n - \frac{\nu(1-\nu)}{2}\Delta_j^2 u^n, \\ u_j^{n+1} &= P_3(u^n) = u_j^n - \nu\Delta_{j-1/2}u^n - \frac{\nu(1-\nu)}{2}\Delta_j^2 u^n + \frac{\nu(1-\nu^2)}{6}\Delta^3_{j-1/2}u^n, \\ u_j^{n+1} &= P_4(u^n) = u_j^n - \nu\Delta_{j-1/2}u^n - \frac{\nu(1-\nu)}{2}\Delta_j^2 u^n + \frac{\nu(1-\nu^2)}{6}\Delta^3_{j-1/2}u^n \\ &\quad + \frac{\nu(1-\nu^2)(1-2\nu)}{24}\Delta_j^4 u^n. \end{aligned} \tag{2.5}$$

Now we first choose $L(u^n) = P_1(u^n)$. Concerning the second order scheme, a natural choice for $S(u^n)$ is $P_2(u^n)$. Unfortunatly, it may lead to poor resolution of discontinuities. In [7] an uniformly second order non-oscillatory scheme is applied for $S(u^n)$. It gives a good resolution of discontiunities at the cost of using 7 point stencil in general case(see Fig 7). We apply the following second order TVD schemes for $S(u^n)$ using 5 point stencil:

$$u_j^{n+1} = S(u^n) = u_j^n - \nu\Delta_{j-1/2}u^n - \frac{\nu(1-\nu)}{2}(\phi_j\Delta_{j+1/2}u^n - \phi_{j-1}\Delta_{j-1/2}u^n), \tag{2.6}$$

where $\phi_j = \phi(r_j)$, $r_j = \frac{\Delta_{j-1/2}u}{\Delta_{j+1/2}u}$ is the flux limiter [9]. The fourth order accurate scheme $P_4(u^n)$ is used for $H(u^n)$.As is well known, the scheme $P_4(u^n)$ is strongly oscillatory (see Fig 4). But the scheme with modification (2.1) gives quite good numerical results. We emphasize that our scheme use only 5 points in the stencil of the scheme. It is well known that the fourth order of accuracy is the highest available order in this case.

Remark. If we use P_3 instead of P_4 then we will have third order uniformly accurate high resolution scheme.

It can be proved in the similar way ([7]) that in the case of scalar conservation laws the resulted scheme preserves the conservative properties and satisfies the entropy condition when Δt,Δx tend to zero. Numerecal examples show the uniformly high accuracy of expected order and high resolution abilities for discontinuities.

3. Numerical Tests. We examine our schemes for the periodic Cauchy problems of equation (2.3) ($u(t, x+1) = u(t, x), 0 \le x \le 2$) with the following initial datum:

Problem 1:

$$u(0,x) = sin(\pi x), \tag{3.1}$$

Problem 2:

$$u(0,x) = \begin{cases} exp(-300(x-1)^2), & \text{if } |x-1| < 1/10 \\ 0, & \text{otherwise} \end{cases} \tag{3.2}$$

Problem 3:

$$u(0,x) = \begin{cases} max(0, 1-(4(x-1))^2) & \text{if } |x-1| < 1/4 \\ 0, & \text{otherwise} \end{cases} \tag{3.3}$$

Problem 4:

$$u(0,x) = \begin{cases} 1, & \text{if } |x-1| < 1/3 \\ 0, & \text{otherwise} \end{cases} \tag{3.4}$$

The results are depicted in Fig 1 - 6. The solid lines mean the exact solutions and the markers are the numerical approximates. We can see the uniform accuracy of expected third and fourth orders and high resolution for discontinuities. The application of our schemes in CFD will be reported elsewhere.

Acknowledgement. The author is indebted to Dr. B.X. Jing for many valuable discussions.

References

[1]. A. Harten, High Resolution Schemes for Hyperbolic Conservation Laws. J. Comput. Phys., 49(1983), pp.357 - 393.

[2]. C.W. Shu, TVB Uniformly High - Order Schemes for Conservation Laws, Math. Comp., 49(1987), pp.105 - 121.

[3]. A. Harten, and S. Osher, Uniformly High Order Accurte Non - Oscillatory Schemes I, SIAM J. Num. Anal., 24(1987), pp.279 - 309.

[4]. P. Lax, Systems of Conservation Laws, Commun. on Pure Appl. Math., vol. 13,(1960), pp.217-237.

[5]. H.M. Wu, and M.L. Yang, SOMS - A Second Order Monotone Scheme for Shock Capturing and Its Application to Navier - Stokes Equations. Lecture Notes, No.141 (1980), pp.442 - 447

[6]. T.Y. Hou, and P.L. Floch, Why Nonconservative Schemes Converge to Wrong Solutions: Error Analysis. Ecole Polytechnique, CMA, No: 255 (1992)

[7]. B.X. Jing, On an Essentially Conservative Schemes for Hyperbolic Conservation Laws. Tech. Report, Computing Center Academia Sinica, (1991)

[8]. S. Osher, Convergence of Generalized MUSCL Schemes, SIAM J. Numer. Anal., vol.22 (1985), 947 - 961.

[9]. P.K. Sweby, High Resolution Schemes Using Flux Limiters for Hyperbolic Conservation Laws, SIAM J. Num. Anal., 21 (1984), pp.995 - 1011

[10]. H.M. Wu, Nonconservative High Resolution Schemes, Tech. Report, Computing Center Academia Sinica, (1992)

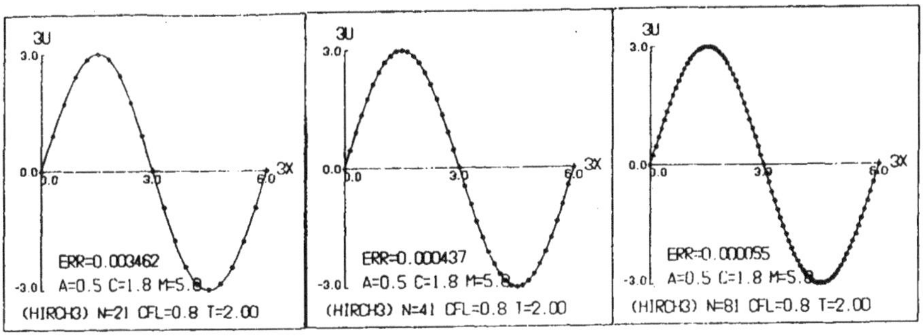

Fig 1. Third order scheme with entropy modification (2.1). Problem 1.

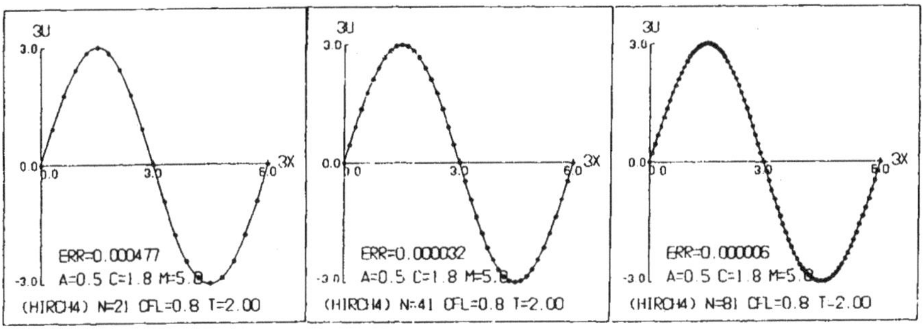

Fig 2. Fourth order scheme with entropy modification (2.1). Problem 1.

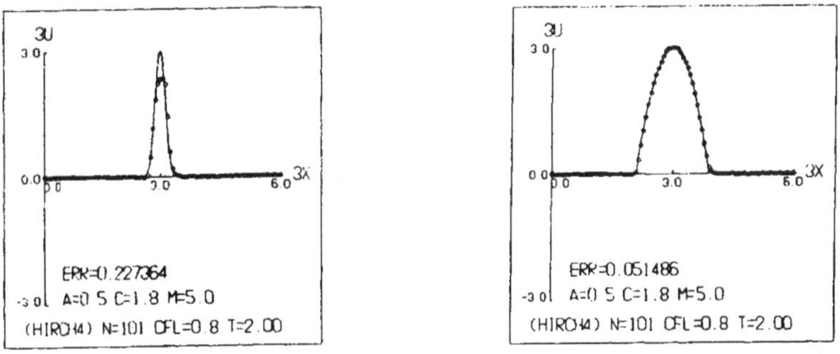

Problem 2.

Problem 3.

Fig 3. Fourth order scheme with entropy modification (2.1).

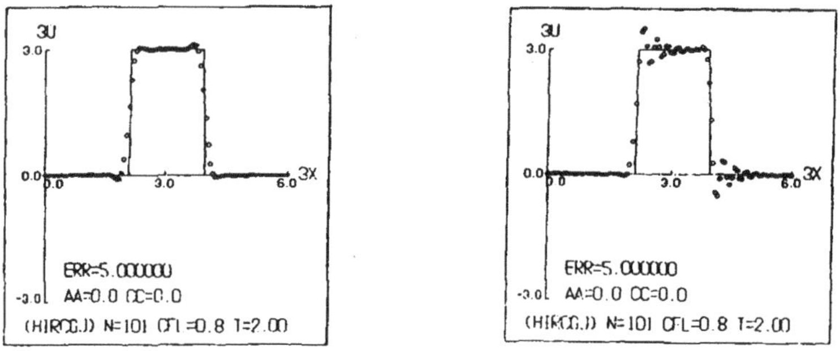

Third order Fourth order

Fig 4. Schemes without entropy modification (2.1) Problem 4.

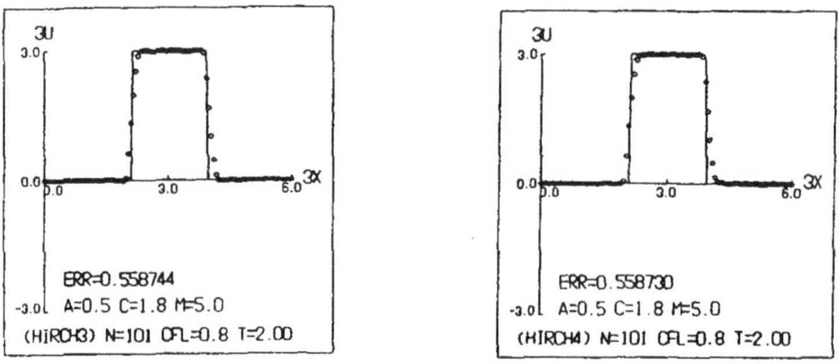

Third order Fourth order

Fig 5. Schemes with entropy modification (2.1) Problem 4.

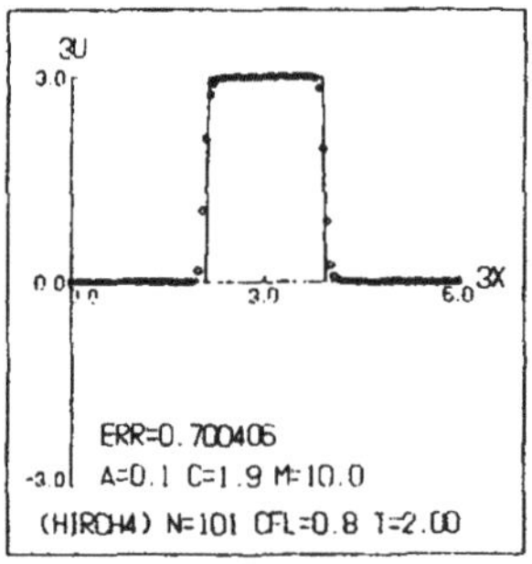

Fig 6. Scheme HIRCH4 with different constants A,C,M

Problem 4.

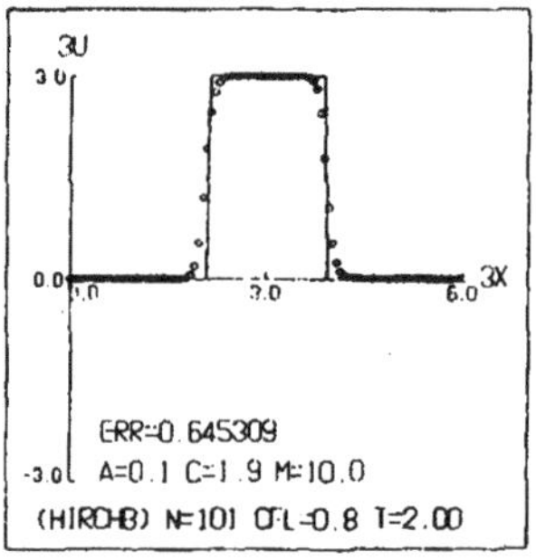

Fig 7. Second order scheme with entropy modification [7]

Problem 4.

A MIXED MULTIPLE TIME SCALE - HARMONIC BALANCE METHOD FOR A FORCED NONLINEAR OSCILLATIONS PROBLEM WITH STEADY STATE SOLUTIONS AND BIFURCATION STUDIES

Julian J. Wu
U.S. Army Research Office, Research Triangle Park, NC, USA
and Duke University, Durham, NC, USA

ABSTRACT

This paper presents applications of a multiple time scale - harmonic balance method to a nonlinear oscillator subjected to a periodic force. The oscillator here is modelled by a single differential equation with linear damping and quadratic and cubic nonlinear springs. The method, which has been published elsewhere, consists of two parts: first, the form of solutions is obtained as a sum of various harmonics, which in turn, are of different orders of magnitudes in terms of a small parameter ε (less than unity); next, the slowly varying coefficients of these harmonics are determined by solving a nonlinear differential equation of a much slower varying variable than the original one. The concept of multiple time scales is used in the first part, and, the harmonic balance approach, used in the second part. Solution formulae in terms of the unknown coefficients of the various harmonics, including a new nonlinear differential equation of a slower varying function than the original unknown variable, are summarized here for the subharmonic, superharmonic and primary resonance cases. Selected numerical results are presented for some specific cases of subharmonic oscillations of order 2, superharmonic of order 1/2 and primary resonance. The results here demonstrate that (1) the present method can yield excellent approximate solutions in comparison with integrations of the original equations, (2) information on bifurcation and stability behaviors in some cases of the problem can also be obtained.

1. INTRODUCTION

The purpose of this paper is to present analytical and numerical results for a forced nonlinear oscillation problem using a "multiple scale - harmonic balance" method (MHB). The oscillation problem can be written in the following equation:

$$d^2u/dt^2 + u + 2\mu(du/dt) + \alpha_2 u^2 + \alpha_3 u^3 = 2f\cos(\Omega t) \qquad (1)$$

Although this equation of forced oscillations, with both quadratic and cubic nonlinearities, has been studied by numerous researchers using analytical, numerical and perturbation methods, its behavior has not been completely understood and the search for further understanding of the problem, in term of steady state solutions, instabilities and bifurcations, etc., is still continuing. Extensive bibliographies can be found in many books on this subject (see Ref.1 and 2, for example). In this paper, it is demonstrated that a new tool can be added for such investigations. This method, MHB, the details of which have been given in previous papers[3,4], is a combination of two quite well known methods: the multiple time scales and the harmonic balance method. Next in Section 2, a summary of the solution formulae of (1), in terms of the transient solutions, the steady state solutions will be given. Three cases will be distinguished: the superharmonic, the subharmonic and the primary resonance. The derivation of these formulae will not be repeated here.

The regular perturbation method is one which often works for a large set of problems as long as certain relevant parameters are "small" in the sense discussed in Ref. 3. To obtain a quite wide range of solutions under the assumption (and "restriction") of "small" parameters, we introduce a parameter ε into (1) and change the dependent variable from u to $\bar{u}$ such that

$$u=\varepsilon\bar{u},\ \mu=\varepsilon\bar{\mu},\ \text{and},\ f=\varepsilon\bar{f},\ \text{or}\ f=\varepsilon^2\bar{f}\ \text{in case of primary resonance} \tag{2}$$

Substituting (2) in (1) and dropping the overbars, we have a new equation to be solved:

$$d^2u/dt^2 + u + 2\varepsilon\mu(du/dt) + \varepsilon\alpha_2 u^2 + \varepsilon^2\alpha_3 u^3 = 2f\cos(\Omega t) \tag{3a}$$

for the super- and subharmonic cases, or,

$$d^2u/dt^2 + u + 2\varepsilon\mu(du/dt) + \varepsilon\alpha_2 u^2 + \varepsilon^2\alpha_3 u^3 = 2\varepsilon f\cos(\Omega t) \tag{3b}$$

for the case of primary resonance. Thus, we consider three cases:

$$\Omega = 1+\varepsilon\sigma,\ \text{the primary resonance case} \tag{4a}$$
$$n\Omega = 1+\varepsilon\sigma,\ \text{the superharmonic resonance of order n} \tag{4b}$$
$$\Omega/n = 1+\varepsilon\sigma,\ \text{the subharmonic resonance of order 1/n} \tag{4c}$$

The reason to separate these cases is due to the fact that the harmonics associated with the solution to each case are usually quite different. Hence the solution formulations are different. Problems formulated in the forms of equations (3a) and (3b) have been analyzed extensively by Nayfeh and associates using a multiple time scale method (see Ref. 6 and 7, for example). The present MHB should provide an independent means of comparison in obtaining solutions to the same problems analyzed. Some limited comparisons for a special example were made in Ref. 3.

2. SOLUTION FORMULAE

We summarize the analytical results obtained for the three cases using MHB. Earlier papers[3,4] should be consulted for the details of the derivations.

2.1 The Case for Primary Resonance. In this case, $\Omega=1+\varepsilon\sigma$. To avoid excessively large solutions due to the primary resonance, the second part for the replacement of f in (2) is used. Following the procedure described in Ref. 3 and 4, a perturbation solution accurate to the first order of ε can be written as

$$u=\varepsilon V_0+2V_{1R}\cos(\Omega t)+2\varepsilon V_{2R}\cos(2\Omega t)+2V_{1I}\sin(\Omega t)+2\varepsilon V_{2I}\sin(2\Omega t) \tag{5}$$

with $\Omega = 1+\varepsilon\sigma$. The nonlinear o.d.e. can be derived for V_1:

$$2i(dV_1/dt)-\varepsilon[f+2(\sigma-i\mu)V_1]+\varepsilon^2[(\sigma-i\mu)f/2-\mu^2V_1+CV_1^2\bar{V}_1]=0 \tag{6}$$

where
$$C=(9\alpha_3-10\alpha_2^2)/3 \tag{7}$$

and an overbar denotes the complex conjugate. The harmonics V_0 and V_2 are given in terms of V_1 by:

$$V_0 = -\alpha_2 V_1 \bar{V}_1, \quad V_3 = \alpha_2 V_1^2 \tag{8}$$

By setting $dV_1/dt=0$ in (7), the steady state dispersion relation can also be written:

$$(\rho/2)^2\{4\mu^2+[2\sigma+\varepsilon(\mu^2-C(\rho/2)^2]^2\}=f^2[(-1+\varepsilon\sigma/2)^2+\varepsilon^2\mu^2/4] \tag{9}$$

where $\rho/2$ is the magnitude of V_1.

2.2 The Case for Superharmonic of Order 2. For the case of superharmonic of order 2, one has:

$$\begin{aligned}u=&\varepsilon V_0+2V_{1R}\cos(\Omega t)+2V_{2R}\cos(2\Omega t)+\varepsilon[2V_{3R}\cos(3\Omega t)+2V_{4R}\cos(4\Omega t)]\\&+2V_{1I}\sin(\Omega t)+2V_{2I}\sin(2\Omega t)+\varepsilon[2V_{3I}\sin(3\Omega t)+2V_{4I}\sin(4\Omega t)]\end{aligned} \tag{10}$$

with $2\Omega = 1+\varepsilon\sigma$. The nonlinear o.d.e. can be derived for V_2:

$$2i(dV_2/dt)-\varepsilon[2(\sigma-i\mu)V_2-C_1f^2]+\varepsilon^2[C_{34}f^2+(C_5f^2-\mu^2)V_2+C_2V_2^2\bar{V}_2]=0 \tag{11}$$

where $C_1=16\alpha_2/9,\ C_2=(9\alpha_3-10\alpha_2^2)/3,\ C_5=(1440\alpha_3-1472\alpha_2^2)/135$ (12a)

$C_{34}=C_3+iC_4,\ C_3=40\sigma\alpha_2/27,\ C_4=-104\mu\alpha_2/27$ (12b)

Once V_2 is obtained from (11), the other harmonics can be determined via the following relations in sequence:

$$V_0 = 2\alpha_2(V_1\bar{V}_1 + V_2\bar{V}_2) \tag{13a}$$

$$V_3=8\alpha_2V_1V_2/5,\ V_4=\alpha_2V_2^2/3,\ V_1=\{12f+8\varepsilon[f(\sigma-2i\mu)-4\alpha_2V_2]\}/9 \tag{13b,c}$$

The steady state dispersion relation can be written as:

$$(\rho/2)^2\{4\mu^2+[2\sigma+\varepsilon(\mu^2-C_2(\rho/2)^2-C_5f^2]^2\}=f^4[(C_1+\varepsilon C_3)^2+\varepsilon^2C_4^2] \tag{14}$$

2.3 The Case for Subharmonic of Order 1/2. For the case of subharmonic of order 1/2, one obtains

$$\begin{aligned}u=&\varepsilon V_0+2V_{1R}\cos(\Omega t)+2V_{2R}\cos(2\Omega t)+\varepsilon[2V_{3R}\cos(3\Omega t)+2V_{4R}\cos(4\Omega t)]\\&+2V_{1I}\sin(\Omega t)+2V_{2I}\sin(2\Omega t)+\varepsilon[2V_{3I}\sin(3\Omega t)+2V_{4I}\sin(4\Omega t)]\end{aligned} \tag{15}$$

with $\Omega = 2(1+\varepsilon\sigma)$. The nonlinear o.d.e. can be derived for V_1:

$$2i(dV_1/dt)-\varepsilon[2(\sigma-i\mu)V_1-C_1f\bar{V}_1]+\varepsilon^2[C_{34}f\bar{V}_1+(C_5f^2-\mu^2)V_1+C_2V_1^2\bar{V}_1]=0 \tag{16}$$

where $C_1=-2\alpha_2/3,\ C_2=(9\alpha_3-10\alpha_2^2)/3,\ C_5=(12\alpha_3-5\alpha_2^2)/18$ (17a)

$C_{34}=C_3+iC_4,\ C_3=11\sigma\alpha_2/9,\ C_4=-8\mu\alpha_2/9$ (17b)

Once V_1 is obtained from (16), the other harmonics can be determined via the following relations:

$$V_0 = -\alpha_2(2V_1V_1 + V_2V_2) \tag{18a}$$

$$V_3=\alpha_2V_1V_2/4,\ V_4=\alpha_2V_2^2/15,\ V_2=\{-3f+4\varepsilon[f(\sigma-i\mu)+3\alpha_2V_1^2]\}/9 \tag{18b}$$

The steady state dispersion relation:

$$\{4\mu^2+[\sigma+\varepsilon(\mu^2-C_2(\rho/2)^2-C_5f^2]^2\}=f^2[(C_1+\varepsilon C_3)^2+\varepsilon^2C_4^2] \tag{19}$$

where now $\rho/2$ is the magnitude of V_1.

It is noted that the so called steady state solutions of (9), (14) and (19) are obtained from the assumed conditions that the slowly varying functions, such as V_0, V_{1R}, V_{1I}, etc., become constant as the time t becomes infinite. These equations will produce true approximate solutions if, (a), such solutions do exist, and, (b), the assumption that the slowly varying functions are constant as t becomes very large is valid. Otherwise, these "steady state" solutions may be meaningless.

3. NUMERICAL EXAMPLES: TRANSIENT, STEADY STATE SOLUTIONS AND BIFURCATIONS

In this section, we shall demonstrate the use of the formulae summarized in the previous section to obtain transient solutions, steady state solutions and information on the occurrence of bifurcations.

3.1 Transient and Steady State Solutions for Primary Resonance Examples. Consider equation (3b), with $\varepsilon = 0.01$, $\alpha_2 = 1.5$, $\alpha_3 = 0.5$, $\mu = 1.0607$, $\Omega = 0.8485$ and $f = 3.0$. This corresponds to a case analyzed in Ref. 8. Since Ω is near unity, we use the formulae associated with the primary resonance. Using the steady state solution formula (9), we obtain $\rho_1 = 2.08137$. The transient solution with initial conditions u(0)=0.95, v(0)=u'(0)=-0.10607, using equations (5)-(8), also produce the same value of ρ_1 for large t (t=400, for example). The comparison of solutions between MHB and directly integrating (3a) is shown in Figure 1, where the solid line represents solution by MHB, and the dashed line, by integrating (3a) directly. The point I in Figure denotes the initial conditions. (These same notation is used in all the phase plots which follow.) The agreement is seen to be excellent.

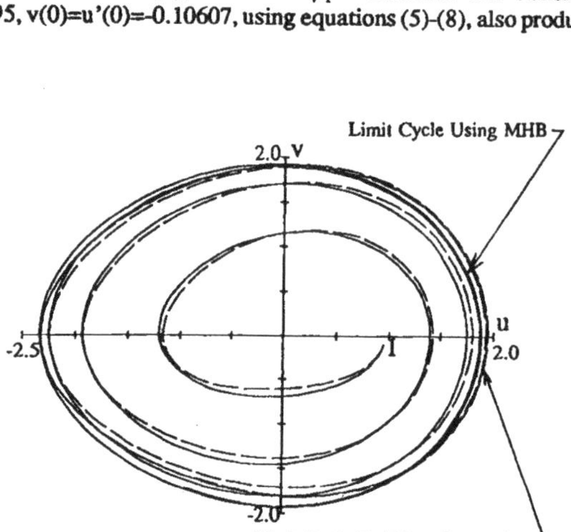

Figure 1. Phase Plot of the Transient and Long Time Solutions for the Case Near Primary Resonance with Damping: ε=0.1, α_2=1.5, α_3=0.5, μ=1.0607, Ω=0.84853, f=3.0, u(0)=0.95, v(0)=0.10607.

3.2 Transient and Steady State Solutions for Superharmonic Examples. In this case, the following set of parameters is used: $\varepsilon = 0.05$, $\alpha_2 = 0.2$, $\alpha_3 = 0.$, $\mu = 5.0$, $\Omega = 0.5$ and $f = 0.5$. Here the steady state equation (14) yields $\rho_2 = 0.02006$. The transient solution with initial conditions u(0) = 1.8192, v(0) = u'(0) = 0.1028 using equations (10)-(13)

also produce the same value of ρ_2 for large t (t=80, for example). The comparison of solutions between MHB and directly integrating (3a) is shown in Figure 2. The "limit cycle" outside is obtained by MHB, the one inside, by direct integration of (3a).

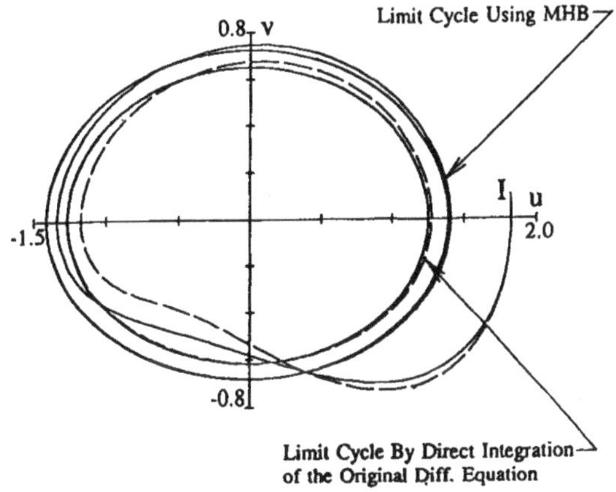

Figure 2. Phase Plot of Transient and the Long Time Solutions for the Case of Superharmonic Resonance with Damping: ε=0.05, α_2=0.2, α_3=0., μ=5.0, Ω=0.5, f=0.5, u(0)=1.8192, v(0)=0.1028.

3.3 Transient Solutions for Subharmonic Examples and Bifurcation Results. As the last numerical example, some results on bifurcation and stability are presented for the subharmonic case. The following set of parameters is used: $\varepsilon = 0.1$, $\alpha_2 = 1.0$, $\alpha_3 = 2.0$, $\mu = 0.4$, $\sigma = 0.6$, and hence $\Omega = 2 + \varepsilon\,\sigma = 2.06$. Using (19) for the steady state solution for the subharmonic case, a solution curve of ρ_1 vs. f is shown in Figure 3 for the range of $1.1 \le f \le 1.9$. This is the same curve presented earlier (Figure 1, curve (b) in Ref. 3, where f, μ and σ were replaced by εf,$\varepsilon\mu$ and $\varepsilon\sigma$ respectively).

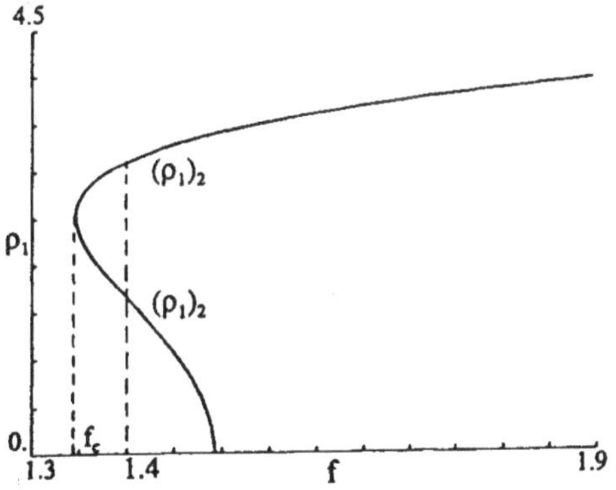

Figure 3. Bifurcation Curve for the Steady State Solutions: ρ_1 vs. f for the Case of Subharmonic Resonance with Damping: $\varepsilon = 0.1$, $\alpha_2 = 1$, $\alpha_3 = 2$, $\mu = 0.4$, $\Omega = 2.06$ and $1.1 \le f \le 1.9$.

Figure 3 shows a bifurcation point at $f_c = 1.346$, where the number of solutions has increased from 1 to 2, not including the trivial solution $\rho_1 = 0$. For $f > f_c$, there are two steady state solutions. At f = 1.4, for example, there are two values for ρ_1: $(\rho_1)_1 = 1.6432$ and $(\rho_1)_2 = 3.0940$. A transient solution will tend to one of these steady state depending on initial conditions. Hence two different cases are computed: Case 1: For u(0) = 0.9588 and v(0) = 0.0273, the value of ρ_1 moves to near the neighborhood of $(\rho_1)_1$, e.g., ρ_1 = 1.8800 at t=400; and, Case 2: With u(0) = 2.5619 and

$v(0) = -0.0436$, on the other hand, ρ_1 converges to $(\rho_1)_2$, e.g., $\rho_1 = 3.0938$ at $t=400$ (with $\Delta t = 0.1$).

The results obtained by MHB are compared with those by directly integrating the original equation (3a) as show in Figure 4, which indicates that the long time silution, for final t of computation, $t_f=400$ ($\Delta t=0.1$), no sharpe limit cycle was reached. A somewhat "diffused" limit cycle obtained using MHB can be observed and is in general agreement with one using direct numerical integration for Case 1. However for Case 2, the solution by numerical integration still converges to the solution for Case 1, rather than to the limit cycle predicted by MHB. This is a numerical indication of an instability for the steady state solution of the original equation corresponding to the MHB steady state solution identified with $(\rho_1)_2$.

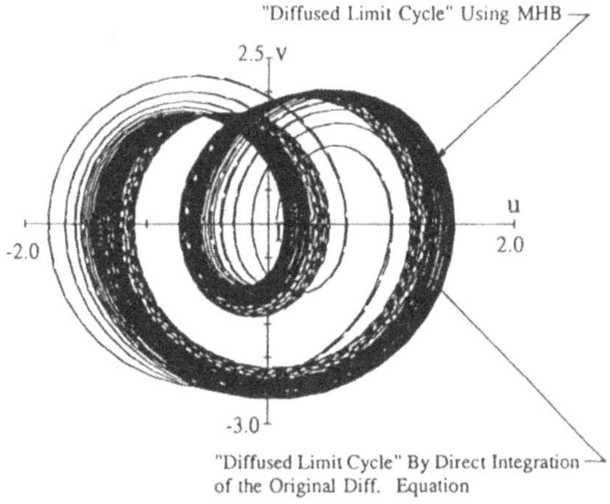

Figure 4. Phase Plot of Transient and the Long Time Solutions for the Case of Subharmonic Resonance with Damping: $\varepsilon=0.1$, $\alpha_2=1.0$, $\alpha_3=2.0$, $\mu=0.4$, $\Omega=2.06$, $f = 1.4$, $u(0)=0.9588, v(0)=0.0273$.

ACKNOWLEDGEMENTS: This work was supported by the US Army Research Office under a staff research program with Duke University.

REFERENCES

1. A. H. Nayfeh and D. T. Mook, ***Nonlinear Oscillations***, (Wiley, New York, 1979).
2. D. W. Jordan and P. Smith, ***Nonlinear Ordinary Differential Equations, Second Ed.***, (Oxford, 1987.
3. J. J. Wu, A generalized harmonic balance method for forced oscillations - the subharmonic cases. To appear in ***Journal of Sound and Vibration***.
4. J. J. Wu, On superharmonic oscillations. ***Computational and Applied Mathematics II: Differential Equations***, W. F. Ames and P. J. van der Houwen, Editors, (Elsevier, 1992), p. 237.
5. B. Noble, Harmonic Balance with Quadratic Nonlinearities. To appear.
6. A. H. Nayfeh, The response of single degree of freedom systems with quadratic and cubic non-linearities to a subharmonic excitation. ***Journal of Sound and Vibration***, 89(4), (1983), pp.457-470.
7. A. H. Nayfeh, Quenching of Primary Resonance by a Superharmonic Resonance. ***Journal of Sound and Vibration***, 89(4), (1984), pp. 363-377.
8. D. W. Jordan and P. Smith, same as Ref. 2, p. 337.

Asymptotic Stability for the Solutions of Difference Schemes

YING LONG-AN
Department of Mathematics, Peking University,
Beijing, 100871, China

Abstract. We present some results on the existence of discrete travelling waves for the hyperbolic conservation laws and the asymptotic stability of the travelling wave, namely, if the initial data is a perturbation of a travelling wave or the superposition of some travelling waves, the solutions of difference schemes may tend to the travelling waves as $t \to +\infty$.

We consider the initial value problem of the system of conservation laws,

$$\frac{\partial u}{\partial t} + \frac{\partial f(u)}{\partial x} = 0,$$
$$u|_{t=0} = u_0(x),$$

where $u = (u_1, \ldots, u_m)$, $f = (f_1, \ldots, f_m)$. We assume that the system is strictly hyperbolic and the characteristic fields are either genuinely nonlinear or linearly degenerate. Let u_j^n be the value of difference solution at point $(j\Delta x, n\Delta t)$. The explicit difference scheme reads

$$u_j^{n+1} = G(u_{j-r}^n, \ldots, u_{j+l}^n). \tag{1}$$

Suppose u_l and u_r satisfy the Rankine–Hugoniot condition and entropy condition, then there is a travelling shock wave

$$u = \phi(x - st),$$

where s is the shock speed and

$$\phi(\xi) = \begin{cases} u_l, & \xi < \xi_0 \\ u_r, & \xi > \xi_0. \end{cases}$$

The first question is; is there a discrete travelling shock wave related to scheme (1) and u_l, u_r ?

The difficulty for defining a discrete travelling wave lies in that if (x_0, t_0) is a grid point, then generally speaking $(x_0 - ns\Delta t, t_0 - n\Delta t)$, n integer, are not grid points, so it is meaningless to say the solution is constant along a line with slope s. So far there are two ways to define a travelling wave.

1. Fine mesh.

Let $\lambda = \Delta t/\Delta x$, $\eta = s\lambda$,

$$\mathcal{L}_\eta = \overline{\{m\eta + n | m, n \in \mathbb{Z}\}}.$$

Consider all grid points $(j\Delta x, n\Delta t)$, where $j \in \mathcal{L}_\eta$. Then ϕ_j is a travelling wave if

$$\phi_{j-\eta} = G(\phi_{j-r}, \ldots, \phi_{j+l}), \qquad j \in \mathcal{L}_\eta.$$

2. Coarse mesh.

If η is a rational number p/q, we consider a difference scheme with respect to a large time step $q\Delta t$,

$$u_j^{nq} = G^q\left(u_{j-qr}^{(n-1)q}, \ldots, u_{j+ql}^{(n-1)q}\right),$$

then a travelling wave is defined by

$$\phi_{j-p} = G^q(\phi_{j-qr}, \ldots, \phi_{j+ql}).$$

Jennings proved the existence of travelling waves for scalar conservation laws, conservative and strictly monotonic scheme, and fine mesh[1]. Majda and Ralston applied central manifold theorem to prove that existence was valid for systems with coarse mesh if a non-resonance condition was satisfied and the wave was weak enough[2]. The non-resonance condition is that for linearized difference scheme

$$v_j^{n+1} = \sum_{i=-r}^{l} C_i(u) v_{j+i}^n,$$

where $\lambda_k(u) = s$, is one eigenvalue of the matrix $\frac{\partial f(u)}{\partial u}$, whenever z is a root of

$$\det\left(1 + z^p \left(\sum_{i=-r}^{l} C_i(u) z^i\right)^q\right) = 0,$$

and $|z| = 1$, then $z = 1$. Smyrlis studied the Lax-Wendroff scheme

$$u_j^{n+1} = u_j^n + \kappa(u_{j+1}^n, u_j^n) - \kappa(u_j^n, u_{j-1}^n),$$

$$\kappa(a, b) = -\frac{\lambda(f(a) + f(b))}{2} + \frac{\lambda^2}{2} A\left(\frac{a+b}{2}\right)(f(a) - f(b)),$$

$$A(u) = f'(u).$$

The existence of stationary shock($s = 0$) was proved[3]. Yu studied the same scheme and proved that there was a continuous function $d(s)$ such that $d(0) = 0$ and if $|u_l - u_r| < d(s)$ then travelling wave failed to exist[4].

The second question is: if a travelling wave exists, is it asymptotically stable ?

In [1] it was proved that if ϕ_j was a travelling wave and u_j^n the perturbed difference solution, then

$$\sum_{j \in \mathcal{L}_n} |u_j^n - \phi_{j-n\eta}| \to 0 (n \to +\infty).$$

The proof was incomplete. Ralston indicated this fact and gave anothor proof in an unpublished paper. Engquist and Osher[5] quoted Ralston's proof, still incomplete. One counter example was given by Zhou[6]. However if we consider coarse mesh instead the fine one, the result is valid and can be proved. In [3] Smyrlis proved asymptotic stability of stationary shocks given by the LW scheme.

The Lax-Friedrichs scheme

$$u_j^{n+1} - \frac{u_{j+1}^n + u_{j-1}^n}{2} + \frac{\lambda}{2}(f(u_{j+1}^n - f(u_{j-1}^n)) = 0 \tag{2}$$

is monotonic for scalar conservation laws, but not strictly monotonic. Chern studied scheme (2) for systems[7]. Using the spectrum method he proved if the initial data was compactly supported, then

$$u_j^n \to 0 \qquad (n \to +\infty).$$

We note that zero is then the travelling wave. Liu and Xin considered a generalized Lax-Friedrichs scheme

$$u_j^{n+1} - u_j^n + \frac{\lambda}{2}\left(f(u_{j+1}^n - f(u_{j-1}^n)\right) = \frac{\alpha}{2}\left(u_{j+1}^n - 2u_j^n + u_{j-1}^n\right),$$

where $0 < \alpha \leq 1$. It is reduced to the LF scheme at the case $\alpha = 1$. Using energy method they proved that if $0 < \alpha < 1$ and

$$\sum_j (u_j^0 - \phi_j) = 0,$$

then $u_j^n - \phi_j^n \to 0$ $(n \to +\infty)$ in the space L^p, $i < p < +\infty$[8]. Ying and Zhou studied the LF scheme (2) for systems. It was proved that if there were two travelling waves ϕ_j and ψ_j such that

$$\sum_{j=odd} (u_j^0 - \phi_j^0) = \sum_{j=even} (u_j^0 - \psi_j^0) = 0,$$

then

$$\sum_{j+n=odd} |u_j^n - \phi_j^n| \to 0, \qquad \sum_{j+n=even} |u_j^n - \psi_j^n| \to 0,$$

as $n \to +\infty$[9].

REFERENCES

[1]. G. Jennings, *Discrete shocks*, Comm. Pure Appl. Math. **27** (1974), 25–37.

[2]. A. Majda and J. Ralston, *Discrete shock profiles for systems of conservation laws*, Comm. Pure Appl. Math. **32** (1979), 445–483.

[3]. Y. S. Smyrlis, *Existence and stability of stationary profiles of the LW scheme*, Comm. Pure Appl. Math. **43** (1990), 508–545.

[4]. S. H. Yu, *Nonexistence of the local travelling profiles for the Lax-Wendroff scheme*, preprint (1992).

[5]. B. Engquist and S. Osher, *One-sided difference approximations for nonlinear conservation laws*, Math. Comp. **36** (1981), 321–351.

[6]. T. Zhou, *The long time asymptotic behavior of the monotonic schemes*, MS thesis, Peking University (1990).

[7]. I. L. Chern, *Large time behavior of solutions of Lax–Friedrichs finite difference equations for hyperbolic systems of conservation laws*, Math. Comp. **56** (1991), 107–118.

[8]. J. G. Liu and Z. P. Xin, *Nonlinear stability of discrete shocks for systems of conservation laws*, preprint (1992).

[9]. L.–a. Ying and T. Zhou, *Long time asymptotic behavior of the Lax–Friedrichs scheme*, J. PDE(to appear).

Paraxial Factorization of Wave Operator and Inverse Problem of Wave Equation[1]

Zhang Guan-quan

(State Key Laboratory of Scientific and Engineering Computing, Computing Center of Chinese Academy of Sciences, Beijing, 100080, China)

Song Hua

(Department of Mathematical Sciences, Rice University, Houston, U.S.A.)

Abstract

Paraxial factorization of two-dimensional wave operator is discussed. The inverse problem of wave equation to recovering the unknown potential is reformulated by coupling the downgoing and upcoming waves. A numerical method of layer-stripping type is proposed.

Introduction

The inverse problem of one-dimensional wave equation have been very well studied. Sufficiently complete results about existence and stability of their solutions are obtained and many numerical methods are proposed. But in the multi-dimensional case, very little is known. The difficulty is lain in the ill-posedness of the problem, caused by evanescent wave. In this article we discuss the factorization of two-dimensional wave operator and the decomposition of wavefield into downgoing and upcoming waves by using so-called square-root operator. Then we reformulate the potential-inversion problem with aide of the system of coupled equations for downgoing and upgoing waves. The original ill-posed inverse problem of wave equation is thus regularized by neglecting evanescent wave. The reformulated potential-inversion problem is a nonlinear initial boundary value problem. The equations governing the downgoing and upcoming waves are integro-differential with some auxiliary differential equations. These equations can be approximated by using quadrature formula and difference schemes. The approximate potential-inversion problem thus obtained can be solved layer-by-layer marched in direction z.

1. Potential-inversion Problem

[1])China State Major Key Project for Basic Researches

The inverse problem considered in this article is the following

$$\begin{cases} \left[\frac{\partial^2}{\partial t^2} - \frac{\partial^2}{\partial x^2} - \frac{\partial^2}{\partial z^2} + v(x,z)\right]u(x,z,t) = 0, \quad \forall x \in (-\infty,\infty), \quad t>0, \ z>0, & (1.1) \\ u(x,0,t) = \delta(t), & (1.2) \\ \frac{\partial u}{\partial z}(x,0,t) = f(x,t), \quad \forall x \in (-\infty,\infty), \quad t>0, & (1.3) \\ u(x,z,t=0) = \frac{\partial u}{\partial t}(x,z,t=0) = 0, & (1.4) \end{cases}$$

where $f(x,t)$ is the boundary response excited by the impulsive line source (1.2).The objective of the problem is to determine the unknown coefficient(potential) $v(x,z)$ in equation (1.1).

The singularity of source (1.2) will be propagated. The generalized solution of (1.1) can be found in the form[1]

$$u(x,z,t) = \delta(t-z) + a(x,z,t)H(t-z), \tag{2}$$

where $H(t)$ is Heaviside function and $a(x,z,t)$ is regular function.Substituting (2) in (1.1) and analysing the singularities .We have

$$\left(\frac{\partial}{\partial t} + \frac{\partial}{\partial z}\right)u(x,z,t=z+0) = \left(\frac{\partial}{\partial t} + \frac{\partial}{\partial z}\right)a(x,z,t=z) = -\frac{v(x,z)}{2}, \tag{3}$$

So

$$u(x,z,t=z+0) = a(x,z,z) = -\int_0^z \frac{v(x,y)}{2}dy. \tag{4}$$

Obviously $u(x,z,t) \equiv 0$ for $t < z$.Therefore we can limit our consideration only in the region $\Omega = \{(x,z,t) : x \in (-\infty,\infty), t > z > 0\}$ for equation (1.1) with conditions (1.2), (1.3), (4) and $v(x,z)$ determined by (3). Unfortunately, even the forward problem (1.1-3) with known potential $v(x,z)$ and one condition of (1.4) ($u(\cdot,t) = 0$, or $\frac{\partial u(\cdot,t=0)}{\partial t} = 0$) is ill-posed.

2. Reformulation of Potential-inversion Problem

Now we factorized the wave operator[2]

$$\frac{\partial^2}{\partial t^2} - \frac{\partial^2}{\partial x^2} - \frac{\partial^2}{\partial z^2} = \left(\Lambda + \frac{\partial}{\partial z}\right)\left(\Lambda - \frac{\partial}{\partial z}\right) = \left(\Lambda - \frac{\partial}{\partial z}\right)\left(\Lambda + \frac{\partial}{\partial z}\right) \tag{5}$$

where Λ is a pseudo-differential operator with the symbol $\lambda(\tau,\xi) = i\tau(1-\xi^2/\tau^2)^{\frac{1}{2}}$.$\Lambda$ is called square-root operator.

In the region $|\xi/\tau| < 1$, we have the identity

$$(1-\xi^2/\tau^2)^{\frac{1}{2}} = 1 - \frac{1}{\pi}\int_{-1}^{1}\sqrt{1-s^2}\frac{\xi^2 ds}{\tau^2 - s^2\xi^2} \tag{6}$$

Then the pseudo-differential operator Λ can be represented in the following form

$$\Lambda = \frac{\partial}{\partial t}[I - \frac{1}{\pi}\int_{-1}^{1}\sqrt{1-s^2}Q(s)ds] = \frac{\partial}{\partial t}[I-R], \tag{7}$$

where I is the identity operator and $Q(s)$ is a pseudo-differential operator of order zero($OP(0)$), satisfying

$$(\frac{\partial^2}{\partial t^2} - s^2\frac{\partial^2}{\partial x^2})Q(s) = \frac{\partial^2}{\partial x^2}. \tag{8}$$

We introduce the upcoming wave and the downgoing wave[2]

$$\begin{cases} \dfrac{\partial U}{\partial t} = \dfrac{1}{2}[\dfrac{\partial}{\partial t}(I-R) + \dfrac{\partial}{\partial z}]u(x,z,t) = \dfrac{1}{2}(\Lambda + \dfrac{\partial}{\partial z})u, \\ \dfrac{\partial D}{\partial t} = \dfrac{1}{2}[\dfrac{\partial}{\partial t}(I-R) - \dfrac{\partial}{\partial z}]u(x,z,t) = \dfrac{1}{2}(\Lambda - \dfrac{\partial}{\partial z})u, \end{cases} \tag{9}$$

from which we have

$$u = H(D+U) \qquad \text{with } H(I-R) = (I-R)H = I, \tag{10}$$

where H is an operator $OP(0)$ with the symbol $h = (1-\xi^2/\tau^2)^{-\frac{1}{2}}$.

It is easy to verified that

$$Hf = \frac{1}{\pi}\int_{-1}^{1}\sqrt{1-s^2}P_f(s)ds + f, \tag{11}$$

where P_f satisfies the following integro-differential equation

$$(\frac{\partial^2}{\partial t^2} - s^2\frac{\partial^2}{\partial x^2})P_f(s) - \frac{\partial^2}{\partial x^2}(\frac{1}{\pi}\int_{-1}^{1}\sqrt{1-\eta^2}P_f(\eta)d\eta) = \frac{\partial^2}{\partial x^2}f. \tag{12}$$

With the aid of factorization (5), the original inverse problem (1.1-4) can be reformulated in the terms of upcoming wave $\frac{\partial U}{\partial t}$ and downgoing wave $\frac{\partial D}{\partial t}$ as follows

$$\begin{cases} [\dfrac{\partial}{\partial t}(I-R) - \dfrac{\partial}{\partial z}]\dfrac{\partial U}{\partial t} + \dfrac{v}{2}H(D+U) = 0, \\ [\dfrac{\partial}{\partial t}(I-R) + \dfrac{\partial}{\partial z}]\dfrac{\partial D}{\partial t} + \dfrac{v}{2}H(D+U) = 0, \end{cases} \quad \forall x \in (-\infty,\infty),\ t > z > 0, \tag{13.1}$$

$$\begin{cases} \dfrac{\partial D(x,z=0,t)}{\partial t} = -\dfrac{1}{2}\dfrac{\partial u(x,z=0,t)}{\partial z} = -\dfrac{1}{2}f(x,z), \quad \forall t > 0, \\ \dfrac{\partial U(x,z=0,t)}{\partial t} = \dfrac{1}{2}\dfrac{\partial u(x,z=0,t)}{\partial z} = \dfrac{1}{2}f(x,z), \quad \forall t > 0, \end{cases} \tag{13.2}$$

$$\begin{cases} D(x,z,t=z+0) = u(x,z,t=z+0) = a(x,z,z) = -\int_0^z \frac{v(x,y)}{2}dy & (13.3.1) \\ \frac{\partial D(x,z,t=z+0)}{\partial t} = \frac{1}{2}(\frac{\partial}{\partial t} - \frac{\partial}{\partial z})u(x,z,t=z+0) = \frac{1}{2}g_D(x,z) & \\ \quad = \frac{1}{2}\int_0^z [\frac{\partial^2}{\partial x^2}a(x,y,y) - v(x,y)a(x,y,y)]dy & (13.3.2) \end{cases}$$

$$\lim_{t\to\infty} U(x,z,t) = \lim_{t\to\infty} \frac{\partial U(x,z,t)}{\partial t} = 0, \tag{13.4}$$

with

$$v(x,z) = -\frac{1}{4}\frac{\partial U(x,z,t=z+0)}{\partial t}. \tag{13.5}$$

Conditions (13.5) (13.3)(13.2) are obtained from conditions (1.2)(1.3) and singularity analysis.Equations (13.1) are obtained directly from equation (1.1), factorization (5) and definitions (9) (10).Condition (13.4) is supplementary in order to limit the behavior of U and $\frac{\partial U}{\partial z}$ for $t \to \infty$, which implicates $v(x,z) \to 0$ for $z \to \infty$.

In equations (13.1)

$$\begin{pmatrix} RU \\ RD \end{pmatrix} = -\frac{1}{\pi}\int_{-1}^{1} \sqrt{1-s^2}\begin{pmatrix} q_U(s) \\ q_D(s) \end{pmatrix} ds, \tag{14}$$

$$\begin{pmatrix} HU \\ HD \end{pmatrix} = \begin{pmatrix} U \\ D \end{pmatrix} + \frac{1}{\pi}\int_{-1}^{1} \sqrt{1-s^2}\begin{pmatrix} P_U(s) \\ P_D(s) \end{pmatrix} ds, \tag{15}$$

where $q_U(s) = Q(s)U$, $q_D(s) = Q(s)D$ and $P_U(s)$, $P_D(s)$ satisfy respectively equation (8) and (12) with conditions

$$q_D(s;x,z,t=z) = \frac{\partial q_D(s;x,z,t=z)}{\partial t} = P_D(s;x,z,t=z) = \frac{\partial P_D(s;x,z,t=z)}{\partial t} = 0, \tag{16}$$

$$\lim_{t\to\infty} q_U(s;x,z,t) = \lim_{t\to\infty} \frac{\partial q_U(s;x,z,t)}{\partial t} = \lim_{t\to\infty} P_U(s;x,z,t) = \lim_{t\to\infty} \frac{\partial P_U(s;x,z,t)}{\partial t} = 0, \tag{17}$$

Conditions (16) is obtained by singularity analysis and condition (17) is supplementary in accordance with (13.4).

It can be proved[2] that the wavefront of equation (1.1) is also that of coupled system (13.1) and the solution of the transport equation of (13.1) can be derived directly from that of (1.1). Indeed factorization (5) is exact for radiation wave $u(x,z,t)$ in which Fourier transform of u satisfies $\hat{u}(\xi,z,\tau) \equiv 0$ for $|\xi/\tau| > 1$.

In contrast with the ill-posed forward problem for original equation (1.1) mentioned above, forward problem (13.1-4) with known $v = 0$ is well-posed[3]. But the well-posedness of inverse problem (13.1-5) remains to prove.

3. Discrete Approximation

Now we set up the discrete counterpart of potential-inversion problem. First we use quadrature formula to evaluate the integrals involved in integral form (7) of square-root operator and auxiliary equation (12) of P_U and P_D. If the Gaussian formula[4] is taken, the integral is approximated as follows

$$\frac{1}{\pi}\int_{-1}^{1}\sqrt{1-s^2}f(s)ds \approx \sum_{k=1}^{n} a_k f(s_k). \tag{18}$$

where

$$s_k = \cos(\frac{k\pi}{n+1}), \quad a_k = \frac{1}{n+1}\sin^2(\frac{k\pi}{n+1}). \tag{19}$$

Then system (13.1) is replaced by the following system

$$\begin{cases} (\frac{\partial}{\partial t} - \frac{\partial}{\partial z})\frac{\partial U}{\partial t} - \frac{\partial^2}{\partial t^2}\sum_{k=1}^{n} a_k q_U(s_k,\cdot) \\ \qquad +\frac{v}{2}[D+U+\sum_{k=1}^{n} a_k(P_D(s_k,\cdot)+P_U(s_k,\cdot))] = 0, \\ (\frac{\partial}{\partial t} + \frac{\partial}{\partial z})\frac{\partial D}{\partial t} - \frac{\partial^2}{\partial t^2}\sum_{k=1}^{n} a_k q_D(s_k,\cdot) \\ \qquad +\frac{v}{2}[D+U+\sum_{k=1}^{n} a_k(P_D(s_k,\cdot)+P_U(s_k,\cdot))] = 0 \end{cases} \tag{13.1'}$$

where q_U,q_D and P_U,P_D satisfy respectively

$$(\frac{\partial^2}{\partial t^2} - s^2\frac{\partial^2}{\partial x^2})q_f(s,\cdot) = \frac{\partial^2 f}{\partial x^2}, \tag{8'}$$

$$(\frac{\partial^2}{\partial t^2} - s^2\frac{\partial^2}{\partial x^2})P_f(s,\cdot) - \frac{\partial^2}{\partial x^2}\sum_{k=1}^{n} a_k P_f(s_k,\cdot) = \frac{\partial^2 f}{\partial x^2}. \tag{12'}$$

With aide of (8′), system (13.1′) can be written in another form

$$\begin{cases} (\frac{\partial}{\partial t} - \frac{\partial}{\partial z})\frac{\partial U}{\partial t} - \frac{1}{2}\frac{\partial^2 U}{\partial x^2} - \frac{\partial^2}{\partial x^2}\sum_{k=1}^{n} a_k s_k^2 q_U(s_k,\cdot) \\ \qquad +\frac{v}{2}[D+U+\sum_{k=1}^{n} a_k(P_D(s_k,\cdot)+P_U(s_k,\cdot))] = 0, \\ (\frac{\partial}{\partial t} + \frac{\partial}{\partial z})\frac{\partial D}{\partial t} - \frac{1}{2}\frac{\partial^2 D}{\partial x^2} - \frac{\partial^2}{\partial x^2}\sum_{k=1}^{n} a_k s_k^2 q_D(s_k,\cdot) \\ \qquad +\frac{v}{2}[D+U+\sum_{k=1}^{n} a_k(P_D(s_k,\cdot)+P_U(s_k,\cdot))] = 0. \end{cases} \tag{13.1''}$$

In numerical experiment n is generally taken to be 1 or 2. For system (13.1″) and auxiliary equations (8′) (12′), the difference schemes,similar to that[5,6] constructed for migration problem, can be used. In numerical solution, D,U and their auxiliary functions q_D,q_U,P_D,P_U are obtained on the march in direction z and the unknown potential $v(x,z)$ is determined layer-by-layer.

Numerical experiment is performed for the laterally varied model $v(x,z) = \bar{v}(x)$ and the dipping model $v(x,z) = \bar{v}(x - ax)$. The numerical results obtained by Song for these models with the synthetic data $f(x,z) = \dfrac{\partial u(x,z=0,t)}{\partial z}$ are satisfactory. The details of numerical procedure and the results of numerical experiment will be reported in another article.

References

[1] R.Courant and D.Hilbert, Methods of Mathematical Physics, **Vol.2**, *John Wiley, New York*, 1962.

[2] Zhang Guanquan, System of coupled equations for upcoming and downgoing waves, *Acta Math. Appl. Sinica*, **Vol. 16**, No.2.

[3] Zhang Guanquan, Square-root Operator and Its applications in numerical solution of some wave problems, *First Chinese-Japan joint seminar on numerical mathematics, Beijing*, 1992.

[4] P.Davis and P.Rabinowitz, Methods of Numerical Integration, *Academic Press, Inc.*, 1984.

[5] Zhang Guanquan, Steep dip finite-difference migration using the system of lower-order partial differential equations, *Acta Geophysica Sinica*, **Vol. 29**, No.3, pp.273-282, 1986.

[6] Zhang Guanquan et al., A new algorithm for finite-difference migration of steep dips, *Geophysics*, **Vol. 53**, No.2, pp.167-175, 1988.

Numerical Schemes for Conservation Laws via Hamilton-Jacobi Equations

L. CORRIAS & R. NATALINI

Istituto per le Applicazioni del Calcolo, "M. Picone"
Viale del Policlinico 137, 00161 Roma, Italy

M. FALCONE

Dipartimento di Matematica, "G. Castelnuovo"
Università di Roma "La Sapienza", P.zza A. Moro
00185 Roma, Italy

ABSTRACT

We present some difference approximation schemes which converge to the entropy solution of a scalar conservation law having a convex flux. The numerical methods described here take their origin from approximation schemes for Hamilton-Jacobi-Bellman equations related to optimal control problems and exhibit several interesting features: the convergence result still holds for quite arbitrary time steps, the main assumption for convergence can be interpreted as a discrete analogue of Oleinik's entropy condition, numerical diffusion around the shocks is very limited.

1. Introduction.

We are interested in the approximation of the entropy weak solution in the sense of Kruzkov [Kr1] of the following scalar conservation law

$$(CL) \qquad \begin{cases} u_t + f(u)_x & = 0 \qquad \text{in } \mathbf{R} \times (0,\infty) \\ u(x,0) & = u_0(x) \quad \text{in } \mathbf{R} \end{cases}$$

The connections between this problem and the theory of generalized solutions (see e.g. [Ca], [CEL], [Li]) to the first order Hamilton-Jacobi equation

$$(HJ) \qquad \begin{cases} v_t + f(v_x) & = 0 \qquad \text{in } \mathbf{R} \times (0,\infty) \\ v(x,0) & = v_0(x) \quad \text{in } \mathbf{R} \end{cases}$$

are known. Roughly speaking, if v is a viscosity solution of (HJ), then $u := v_x$ is an entropy solution of (CL). This relation has been used also for numerical purposes in order to derive schemes for (HJ) from the huge collection of methods for (CL). Several authors have used this technique, e.g. Kruzkov in [Kr2] and Crandall and Lions in [CL] have studied numerical methods derived from first order schemes for (CL), Osher and Sethian in [OS] have obtained schemes for (HJ) starting from ENO schemes.

We also use this relation for numerical purposes but in the opposite direction. The numerical approximation of (HJ) has the advantage that we deal with more regular solutions (typically they are locally Lipschitz continuous) and we make the derivative in x at the end. Another advantage is that the expressions of the schemes for (HJ) are relatively simple in comparison with their analogous for (CL). In particular we will apply to (HJ) numerical methods developed for the Hamilton-Jacobi-Bellman equation related to a finite horizon optimal control problem (see [F], [FG] and references therein).

An interesting feature of these schemes is that they do not work on a fixed number of nodes, since the stencil depend on the ratio $\Delta t/\Delta x$ which can be non constant. Choosing a large time step will only increase the width of the stencil in such a way that the domain of dependence of the conservation law will be always contained in the numerical domain of dependence. Naturally this will require a larger number of operations at each time step but the global CPU time necessary to obtain accurate results is even lower than that needed by some classical schemes. We refer to [Ha1] and [Le] for the study and the numerical experiments related to other difference schemes with large time-step. We observe also that these schemes can be interpreted as a discrete version of Lax-Oleinik-Hopf's representation formulas (see [La], [Ol], [Ho], and [Li]).

In conclusion the HJ-schemes produce accurate approximations both for smooth and for discontinuous solutions. Detailed proofs as well as a complete account about numerical experiments will be found in [CFN].

2. A general convergence result.

Let Δx and Δt be the mesh size and v_j^n be the approximation of the solution v of (HJ) at the grid point $(j\Delta x,\ n\Delta t)$, given by a general finite difference scheme

$$v_j^{n+1} = F(v_{j-p}^n, \ldots, v_{j+p}^n).$$

Define the approximation u_j^n of the solution u of (CL), for example as

$$u_j^n := \frac{v_{j+1}^n - v_j^n}{\Delta x}.$$

The following result still holds true using centered or backward differences.

For Δx and Δt fixed, the approximate solution is given by

$$v_\Delta(x,t) := v_j^n \ , \quad \text{for} \quad (x,t) \in I_{j,n}$$

where, for $j \in \mathbb{Z}$, $n \in \{0, \ldots, N\}$

$$I_{j,n} := \left[\left(j - \frac{1}{2}\right) \Delta x, \left(j + \frac{1}{2}\right) \Delta x \right) \times [n\Delta t, \ (n+1)\Delta t)$$

and analougously for $u_\Delta(x,t)$.

THEOREM 2.1. *Assume that for any $t \in (0, T)$, the sequence v_Δ converges in $L^\infty(\mathbb{R})$ to the viscosity solution of (HJ), as Δx and Δt go to zero, and the following properties are verified:*

(i) *there exists a constant C_1 such that for any $j \in \mathbb{Z}$ and for any integer $n \in \{0, \ldots, N\}$*

$$\left| \frac{v_{j+1}^n - v_j^n}{\Delta x} \right| \leq C_1 \ ; \tag{2.1}$$

(ii) *there exists a constant C_2 such that for any $j \in \mathbb{Z}$ and for any integer $n \in \{0, \ldots, N\}$*

$$\frac{v_{j+1}^n - 2v_j^n + v_{j-1}^n}{\Delta x^2} \leq C_2 \ . \tag{2.2}$$

Then for any $t \in (0, T)$ the sequence u_Δ converges in $L^1_{\text{loc}}(\mathbb{R})$ to the entropy solution u of (CL).

Remark 2.2. We notice that the inequality (2.2) is a discrete analogue of the celebrated Oleinik's entropy condition [Ol] and that for compact support initial data for (CL), the condition (2.2) implies (2.1). Moreover it is straightforward to show that the schemes satisfying (2.1) and (2.2) are locally TVB. Nevertheless we cannot use the Harten's theorem [Ha2] since, in general, they are not in conservation form.

3. HJ-Schemes.

As we mentioned in the introduction we can construct schemes satisfying the assumptions of the general convergence theorem starting from the schemes which have been studied to obtain an approximation of the value function of a finite horizon optimal control problem, i.e. the unique viscosity solution of an Hamilton-Jacobi equation. Here we just mention for future reference that the following estimate holds true

$$|w(jh, nk) - w_j^n| \leq C_1 k^{\frac{1}{2}} + C_2 h + C_3 \frac{h}{k} \tag{3.1}$$

where w is the value function and w_j^n is the approximate solution obtained by applying the above mentioned scheme (see [CFN], [F], [FG]). The corresponding scheme for (HJ) can be written as

$$\begin{cases} v_j^n = \inf_{a \in \mathbf{R}} \{v^{n-1}(jh - ka) + kf^*(a)\} \\ v_j^0 = v_0(jh) \end{cases} \tag{3.2}$$

where $v^n(x) := v(x, nk)$ and f^* is the Legendre-Fenchel transform of f.

Formula (3.2) does not give a standard finite difference scheme since in the right term appears $v^{n-1}(jh - ka)$, that is v^{n-1} computed on points which in general are *not* grid points. If this is the case, we replace $v^{n-1}(jh - ka)$ with the convex combination of v_q^{n-1} and v_{q+1}^{n-1} where $q \in \mathbf{Z}$ is such that $(jh - ka) \in [qh, (q+1)h]$.

Two difficulties arise now in (3.2). The first one is related to the evaluation of the infimum since we have to solve an extremum problem over an unbounded domain and the second one is related to the computatiom of f^*. Anyway, under suitable hypothesis on f ([CFN]), it is possible to restrain the infimum in (3.2) on a bounded set of $\mathbf{R}$ while an algorithm for the computation of the Legendre trasform has been proposed in [Br] and precise error estimates (in L^∞) has been established in [Co].

Now we want to prove that the HJ-scheme (3.2) satisfies the assumptions of the general convergence theorem if the infimum in (3.2) is achived in a set $A := [-M, M]$. Let $a \in [-M, M]$, then the point

$$(jh - ka) \in [(j+i)h,\ (j+i+1)h) =: I_{ji}$$

if and only if

$$a \in \left(-(i+1)\frac{h}{k},\ -i\frac{h}{k}\right) \cap [-M, M] =: A_i$$

where $p := \left[\frac{k}{h} M\right] + 1$ and $i \in P := \{-p, \ldots, p-1\}$. Then the scheme can be written as

$$v_j^{n+1} = \min_{i \in P} \{v_{j,i}^{n+1}\} \tag{3.3}$$

where

$$(3.4)\qquad v_{j,i}^{n+1} := \min_{a\in A_i}\left\{\left(1+i+\frac{k}{h}a\right)v_{j+i}^n - \left(i+\frac{k}{h}a\right)v_{j+i+1}^n + kf^*(a)\right\}\ .$$

THEOREM 3.1. *Assume that v_0 satisfies (2.1) and (2.2). Then the scheme (3.3)-(3.4) satisfies assumptions (i) and (ii) of Theorem 2.1.*

Proof. We will show only that the scheme satisfies (i). The proof of (ii) is analougous.

Let $v_{j+1}^n = v_{j+1,i_1}^n$ and $v_j^n = v_{j,i_2}^n$ where $i_1, i_2 \in P$. We have

$$(3.5)\qquad v_{j+1}^n - v_j^n = v_{j+1,i_1}^n - v_{j,i_2}^n \le v_{j+1,i_2}^n - v_{j,i_2}^n\ .$$

Assume a_1 and a_2 be the points in A_{i_2} where the minimum in (3.4) is achieved, respectively for v_{j+1,i_2}^n and v_{j,i_2}^n. Then (3.5) becomes

$$\begin{aligned}
v_{j+1}^n - v_j^n \le& \left(1+i_2+\frac{k}{h}a_1\right)v_{j+1+i_2}^{n-1} - \left(i_2+\frac{k}{h}a_1\right)v_{j+1+i_2+1}^{n-1} + kf^*(a_1)\\
&-\left(1+i_2+\frac{k}{h}a_2\right)v_{j+i_2}^{n-1} + \left(i_2+\frac{k}{h}a_2\right)v_{j+i_2+1}^{n-1} - kf^*(a_2)\\
\le& \left(1+i_2+\frac{k}{h}a_2\right)\left(v_{j+1+i_2}^{n-1} - v_{j+i_2}^{n-1}\right) \quad - \left(i_2+\frac{k}{h}a_2\right)\left(v_{j+1+i_2+1}^{n-1} - v_{j+i_2+1}^{n-1}\right)\\
\le& \max\{(v_{j+1+i_2}^{n-1} - v_{j+i_2}^{n-1}),\ (v_{j+2+i_2}^{n-1} - v_{j+i_2+1}^{n-1})\}\ .
\end{aligned}$$

In the same way we get

$$v_{j+1}^n - v_j^n \ge \min\{(v_{j+1+i_1}^{n-1} - v_{j+i_1}^{n-1}),\ (v_{j+2+i_1}^{n-1} - v_{j+i_1+1}^{n-1})\}$$

and by induction on the initial vector $\{v_j^0\}_{j\in\mathbf{Z}}$, (i) is proved. ∎

Theorems 2.1 and 3.1 give a convergence result when the original scheme for (HJ) converges. So in particular, by using (3.1), when $h = o(k)$. Anyway, if (3.1) is not satisfied but the ratio $\frac{k}{h}$ is constant, the scheme (3.2) written for (CL) is in conservation form and consistent and its convergence is guaranted by Remark 2.2. The scheme is even a three points monotone scheme if $\frac{k}{h}M < 1$ and $\frac{k}{h}$ is small enough.

In the general case when $\frac{k}{h}$ is non constant, the scheme obtained from (3.2), is not monotone and not in conservation form, according to the classical Lax-Wendroff definition. In fact the number of points involved in the scheme varies according to the ratio $\frac{k}{h}$. However, it is interesting to notice that the CFL condition is not violated since the domain of dependence of the finite difference method include the domain of dependence of the differential operator.

REFERENCES

[Br] Y. BRENIER, *Un Algorithme Rapide pour le Calcul de Transformées de Legendre-Fenchel Discrètes*, C.R. Acad. Sci. Paris, **308** Sèr. I, (1989), 587-589.

[BE] M. BARDI, L. EVANS, *On Hopf's Formulas for Solutions of Hamilton-Jacobi Equations*, Nonlin. Anal. TMA, **8** (1984), 1373–1381.

[Ca] V. CASELLES, *Scalar Conservation Laws and Hamilton-Jacobi Equations in One-Space Variable*, Nonlin. Anal. TMA, **18** (1992), 461-469.

[Co] L. CORRIAS, *Transformées de Legendre-Fenchel Discrètes et Applicationes*, Memoire de D.E.A., Laboratoire d'Analyse Numérique, Université Pierre et Marie Curie, 1992.

[CFN] L. CORRIAS, M. FALCONE, R. NATALINI , *Numerical Schemes for Conservation Laws via Hamilton- Jacobi Equations*, Quaderno IAC n. 2, 1992.

[CEL] M.G. CRANDALL, L. EVANS, P.L. LIONS, *Some Properties of Viscosity Solutions of Hamilton-Jacobi Equations*, Trans. Amer. Math. Soc. **282** (1984), 487–502.

[CL] M.G. CRANDALL, P.L. LIONS, *Two Approximations of Solutions of Hamilton-Jacobi Equations*, Math. Comp. **43** (1984), 1-19.

[F] M. FALCONE, *A numerical approach to the infinite horizon problem*, Appl. Math, Optim, 15, (1987),1-13.

[FG] M. FALCONE, T.GIORGI, *An Approximation Scheme for Evolutive Hamilton-Jacobi Equations*, in preparation.

[Ha1] A. HARTEN, *On a large Time-Step High Resolution Scheme*, Math. Comp., **46** (1986), 379-400.

[Ha2] A. HARTEN, *High Resolution Schemes for Hyperbolic Conservation Laws*, J. Comp. Phys., **49** (1983), 357-393.

[Ho] E. HOPF, *Generalized solutions of non-linear equations of first-order*, J. Math. and Mech., **14** (1965), 951-973.

[Kr1] S.N. KRUŽKOV, *First Order Quasilinear Equations in Several Independent Variables*, Mat. Sb. **81** (1970), 228–255. English transl.: Math. USSR Sb. **10** (1970), 217–243.

[Kr2] S.N. KRUŽKOV, *The Method of Finite Differences for a First-Order Non-Linear Equation with many independent Variables*, Zh. Vychisl. Mat. Fiz., **5** (1966), 884-894.

[La] P. D. LAX, *Hyperbolic Systems of Conservation Laws and the Mathematical Theory of Shock Waves*, SIAM Reg. Conf. Series, Lectures in Applied Math., 11, 1970, 1-47.

[Le] R.J. LEVEQUE, *A Large Time Step Generalization of Godounov's Method for Systems of Conservation Laws*, SIAM J. Numer. Anal., **22** (1985), 1051-1073.

[Li] P.L. LIONS, *Generalized solutions of Hamilton-Jacobi Equations*, Pitman, London, 1982.

[Ol] O.A. OLEĬNIK, *Uniqueness and Stability of the Generalized solution of the Cauchy Problem for a Quasi-Linear Equation*, Uspehi Mat. Nauk, **14**, 1959, 165-170; English transl.: Amer. Math. Soc. Transl., series 2, 33, 1963, 285-290.

[OS] S. OSHER, J.A. SETHIAN, *Fronts Propagating with Curvature-Dependent Speed: Algorithms on Hamilton-Jacobi Formulations*, J. Comp. Phys., **79** (1988), 12-49.

Construction and Application of Higher Order Difference Schemes for Ordinary Differential Equations*

QIN MENG-ZHAO and ZHU WENJIE
State Key Laboratory of Scientific and Engineering Computing
Computing Center, Academia Sinica, Beijing 100080, P.R. China

Abstract

In this paper, we develop a new technique called **multiplicative extrapolation method** which constructs higher order difference schemes for ordinary differential equations by composing lower order ones. We will discuss the stability of the schemes constructed from the trapezoid rule and the centered Euler method. The application of the constructed schemes to ordinary differential equations is obvious and direct, so in this paper, we will apply the constructed scheme to wave and heat equations to show that these schemes are also useful for solving partial differential equations. The multiplicative extrapolation method presented in this paper has a big advantage over additive extrapolation method because it preserves group property of the original schemes, that means if this method is used to construct higher order schemes from lower order symplectic schemes, the higher order ones are also symplectic and if it is applied to volume-preserving schemes, the constructed higher order scheme will also be volume-preserving.

§1. Summary of the Construction Process

We know every one-step difference scheme for ordinary differential equation system $y' = f(y)$ can be formally written as $y_{n+1} = s(\tau)y_n$, where $s(\tau)$ is the integrator corresponding to the difference scheme, and τ is the step-size. Notice that $s(\tau)$ is at least of order 1 since it is a integrator.

An integrator $s^*(\tau)$ is called to be the adjoint integrator of $s(\tau)$, if it satisfies $s^*(-\tau)s(\tau) = I$ and $s(\tau)s^*(-\tau) = I$. We call an integrator $s(\tau)$ is self-adjoint, if $s^*(\tau) = s(\tau)$. We proved that for any integrator $s(\tau)$, $s^*(\tau/2)s(\tau/2)$ (or $s(\tau/2)s^*(\tau/2)$) is a self-adjoint integrator and the symmetric composition $s_1(c_1\tau)s_2(c_2\tau)s_1(c_1\tau)$ of self-adjoint integrators $s_1(\tau)$, $s_2(\tau)$ with $2c_1 + c_2 = 1$ is also a self-adjoint integrator.

When we study the order of some scheme, we often use the Taylor series, but sometimes it is very tedious to expand the Taylor series of a scheme. However, there is an other method: Lie series, it's the method we use in this paper. Since we need not compute out the exact terms of the "Lie series" of a scheme, we just use the form of them, the deduction then becomes simplified.

Let $L_f = f^T D = \sum_{i=1}^n f_i \frac{\partial}{\partial y_i}$ be a first order differential operator. A Lie series is the exponential of a first order linear differential operator $e^{tL_f} = \sum_{n=0}^{\infty} t^n L_f^n/n!$. From the properties of the Lie series, we know that the solution of $y' = f(y)$ can be represented in the form $y(t) = e^{tL_f}y(0)$. If $s(\tau)$ satisfies group property in τ, it

*Supported by National Natural Science Foundation of China.

will be the phase flow of some autonomous differential system, say, $dx/d\tau = f(x)$. However, in our problem $s(\tau)$ is just a one parameter family without group property, so there just exists a formal vector field $f^\tau(x)$, which defines the formal autonomous system $dx/dt = f^\tau(x)$ with the formal phase flow depending on two parameters τ, t: $e^{tL_{f^\tau}}$. The diagonal phase flow operator $e^{tL_{f^\tau}}|_{t=\tau} = e^{\tau L_{f^\tau}}$ is just $s(\tau)$. Since $f^\tau(x)$ is a formal vector field, it is a formal power series in τ. Thus the exponential representation of $s(\tau)$ will be in the form

$$s(\tau) = \exp(\tau A + \tau^2 B + \tau^3 C + \tau^4 D + \tau^5 E + \cdots) \tag{1}$$

and the sequence $\tau A + \tau^2 B + \tau^3 C + \tau^4 D + \tau^5 E + \cdots$, where $A, B, C, D, E, \cdots$ are first order differential operators, may not be convergent,

Use the exponential representation of the integrator and BCH formula, we can get

Theorem 1.1. *Every self-adjoint integrator has an even order of accuracy.*

Theorem 1.2. *Let $s(\tau)$ to be a self-adjoint integrator of order 2n, then the integrator $s(c_1\tau)s(c_2\tau)s(c_1\tau)$ with c_1, c_2 satisfying $2c_1^{2n+1} + c_2^{2n+1} = 0,\ 2c_1 + c_2 = 1$ is self-adjoint and of order $2n+2$.*

Use Theorem 1.2, we can construct schemes of arbitrary even order. Since for any integrator $s(\tau)$, $s_{2n}(\tau) = s(\tau/2)s^*(\tau/2)$ will be an self-adjoint integrator of even order, say, of order $2n$, thus $s_{2n+2}(\tau) = s_{2n}(c_1\tau)s_{2n}(c_2\tau)s_{2n}(c_1\tau)$ will be self-adjoint and of order $2n+2$; and this construction process can be continued to get schemes of arbitrary even order.

H. Yoshida[2] and M. Suzuki[9] get the same result for symplectic explicit integrators used to solve separable Hamiltonian systems. The result we get here can be applied to non-Hamiltonian systems and non-symplectic integrators and is based on their work.

Lets study two useful examples of application of our conclusion. We know the trapezoid method and centered Euler method are self-adjoint and of order 2, then scheme

$$\begin{cases} y_{n+\frac{1}{3}} = y_n + \dfrac{1}{2(2-2^{\frac{1}{3}})}\tau(f(y_n) + f(y_{n+\frac{1}{3}})) \\ y_{n+\frac{2}{3}} = y_{n+\frac{1}{3}} + \dfrac{-2^{\frac{1}{3}}}{2(2-2^{\frac{1}{3}})}\tau(f(y_{n+\frac{1}{3}}) + f(y_{n+\frac{2}{3}})) \\ y_{n+1} = y_{n+\frac{2}{3}} + \dfrac{1}{2(2-2^{\frac{1}{3}})}\tau(f(y_{n+\frac{2}{3}}) + f(y_{n+1})) \end{cases} \tag{2}$$

and

$$\begin{cases} y_{n+\frac{1}{3}} = y_n + \dfrac{1}{(2-2^{\frac{1}{3}})}\tau f(\dfrac{y_n + y_{n+\frac{1}{3}}}{2}) \\ y_{n+\frac{2}{3}} = y_{n+\frac{1}{3}} + \dfrac{-2^{\frac{1}{3}}}{(2-2^{\frac{1}{3}})}\tau f(\dfrac{y_{n+\frac{1}{3}} + y_{n+\frac{2}{3}}}{2}) \\ y_{n+1} = y_{n+\frac{2}{3}} + \dfrac{1}{(2-2^{\frac{1}{3}})}\tau f(\dfrac{y_{n+\frac{2}{3}} + y_{n+1}}{2}) \end{cases} \tag{3}$$

are schemes of order 4 by Theorem 2.

We point out that the Gauss-Legendre methods are self-adjoint, and they are also multistage, one-step methods, so the construction method in this section can also be applied to them as to the trapezoid method and the centered Euler method in the upper examples.

The symmetry or self-adjiont property of a scheme is based on its time reversibility. The time reversibility enables us to use the multiplicative extrapolation method to construct higher order schemes from lower order ones without the loss of the group property of them.

§2. Stability of the Constructed Schemes

We just study the stability of scheme (2), since the stability of (2) and (3) are the same. First, notice that (2) is not A-stable, though the trapezoid method is. To show this, we apply scheme (2) to the test equation

$$y' = \lambda y, \quad y(0) = y_0, \quad \lambda \in C, \quad \mathrm{Re}\,\lambda < 0. \tag{4}$$

which yields

$$y_{n+1} = \frac{(1+c_1 z)(1+c_2 z)(1+c_1 z)}{(1-c_1 z)(1-c_2 z)(1-c_1 z)} y_n \tag{5}$$

where $\lambda\tau/2 = z, z \in C$ and $c_1 = 1/(2-2^{1/3}), c_2 = -2^{1/3}/(2-2^{1/3})$.

Definition 2.1. *The stable region R of scheme (1) is*

$$R = \left\{ z \in C | V(z) = \left| \frac{(1+c_1 z)(1+c_2 z)(1+c_1 z)}{(1-c_1 z)(1-c_2 z)(1-c_1 z)} \right| < 1. \quad \mathrm{Re}\, z < 0 \right\}. \tag{6}$$

Obviously, when $z \to 1/c_2 (< 0)$, $V(z) \to \infty$, which means scheme (2) can not be stable in some adjacent region of $1/c_2$. Thus we get

Theorem 2.2. *Scheme (1) is not A-stable.*

Since scheme (2) is not A-stable, we should figure out the stable region of it. To do this, we first study the roots of the equation $V(z) = 1$ which is equivalent to

$$c_1^2 c_2 (1+e^{i\theta}) z^3 + (2c_1c_2 + c_1^2)(1-e^{i\theta}) z^2 + (2c_1 + c_2)(1+e^{i\theta}) z + (1-e^{i\theta}) = 0, \quad 0 \le \theta < 2\pi \tag{7}$$

By some simple computation, we can get the following theorem

Theorem 2.3. *The three roots of polynomial (7) are in the form $z_1 = -x + yi, z_2 = x + yi, z_3 = wi$, where x, y, w are all real.*

Theorem 2.3 tells us that there is a root of (7) on the imaginary axis, and the two other roots located symmetrically with respect to the imaginary axis. Thus there is only one root on the left open semi-plane. Computation shows these roots form a closed curve S when θ changing from 0 to 2π.

Since in S, $V(z) > 1$ and when $z \to \infty$, $V(z) \to 1$, we get the conclusion of this section

Theorem 2.4 *The stable region of scheme (2) is*

$$R = \left\{ z \in C | z \text{ outside } S \text{ and } \mathrm{Re}\, z < 0 \right\}.$$

We see though scheme (2) is not A-stable, the stable region of it is infinite. The unstable region approximates a very small disk with the maximal "diameter" only about 0.04 and near -1.18 on the real axis. For every definite λ, we can choose some special step length τ, so that $\lambda\tau$ will not fall into S. Thus we know scheme (2) and (3) are still useful for solving stiff problems[7].

§3. Application to Partial Differential Equations

We know when solving partial differential equations, there are several methods such as spectral methods and finite difference methods which can be used to achieve high order accuracy in the space direction, while high order accuracy in the time direction is not so easy to get. So it is obvious that the overall accuracy is often influenced strongly by the relatively unsatisfactory approximation in the time direction. Though the self-adjoint schemes (also called symmetric schemes or reversible schemes) are well known, such as the composed Strang scheme which is of order two, the advantage of these schemes which can be used to construct higher order schemes is long neglected. In this section, we use scheme (2) to solve two kinds of partial differential equations in order to show that the technique introduced in section 1 can be used to overcome the deficiency in the time direction, since theoretically, we can construction schemes of arbitrary even order in the time direction.

Let's first consider the following one dimensional first order wave equation

$$\begin{aligned} u_t + u_x &= 0, \\ u(x,0) &= f(x), \end{aligned} \qquad 0 \le x \le 2\pi \tag{8}$$

with periodic boundary conditions $u(0,t) = u(2\pi, t)$. Since collocation, Galerkin and tau methods are identical in the absence of essential boundary conditions, we just analyze the Fourier-collocation or pseudospectral method[6].

We introduce the collocation points $x_n = 2\pi n/2N$ $(n = 0, \cdots, 2N-1)$ and the vector notation $\mathbf{u} = (u_0, \cdots, u_{2N-1})$ where $u_n = u(x_n, t)$. The collocation equations that approximate (8) is as the followings

$$\frac{\partial \mathbf{u}}{\partial t} = C^{-1} D C \mathbf{u}, \tag{9}$$

where C and D are $2N \times 2N$ matrices whose entries are $C_{kl} = \exp[i(k - N)x_l]/\sqrt{2N}$ $D_{kl} = -k'\delta_{kl}$, where $k' = k - N$ $(1 \le k \le 2N-1)$ and $k' = 0$ if $k = 0$. Since $\|\exp[C^{-1}DC]t\| = 1$, we know the Fourier-collocation method is stable for (10).

Use scheme (2) and Crank-Nicolson scheme to solve (9) with initial value $f(x) = \sin x$ and compare the numerical results with the exact solution $u(x,t) = \sin(x-t)$, we can find scheme (2) is actually more accurate than the Crank-Nicolson scheme. Numerical results are given out in another paper[8]. Here, the results are plotted out in the **Figure 1** in which the ranges of x and t are $[0, 2\pi]$ and $[0, 10]$ respectively. There are 25 curves in **Figure 1**, each curve represents the values of $u(x,t)$ on different collocation points at the same time. The number of collocation points used here is 24(i.e., $N = 12$), and we take step-size $\Delta t = 0.1$.

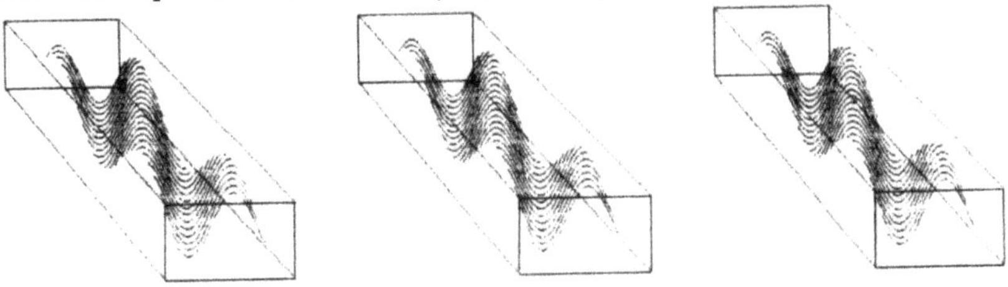

Crank Nicolson Scheme   Scheme (2)   Exact Solution

Figure 1

We now turn to the second order heat equation

$$\frac{\partial u(x,t)}{\partial t} = \frac{\partial^2 u(x,t)}{\partial x^2}, \qquad 0 < x < \pi, \ t \geq 0, \tag{10}$$

with boundary condition $u(0,t) = u(\pi,t) = 0, \quad t > 0$, and initial condition $u(x,0) = f(x), \quad 0 \leq x \leq \pi$. A spectral approximation to (10) is

$$\mathrm{u}(x,t) = \sum_{n=1}^{N} a_n(t) \sin nx \tag{11}$$

$$da_n/dt = -n^2 a_n, \qquad n = 1, \cdots, N. \tag{12}$$

For the initial conditions of a_n, we have $a_n(0) = \frac{2}{N+1} \sum_{j=1}^{N} f(x_j) \sin \frac{\pi j n}{N+1}, \quad n = 1, \cdots, N$, if we let $x_j = \pi j/(N+1)$ $(j = 1, \cdots, N)$ be the collocation points.

The spectral approximation (11)-(12) to (10) is an exceedingly good approximation for any $t > 0$ as $N \to \infty$. In fact, the error $u(x,t) - u_N(x,t)$ goes to zero more rapidly than $e^{-N^2 t}$ as $N \to \infty$ for any $t > 0$.

We now use scheme (1) and trapezoid scheme to solve the evolution equation (12) with the initial condition $f(x) = \sin x$ and then get the solution of (10) from (11). In the computation, we take $N = 24$ and $\Delta t = 0.1$. Here, the results are plotted out in the **Figure 2** in which the ranges of x and t are $[0, \pi]$ and $[0, 5]$ respectively. There are 50 curves in the **Figure 2**, each curve denotes the values of $u(x,t)$ on different collocation points at the same time. In the computation, we take $N = 24$ and $\Delta t = 0.1$.

We have also applied the scheme (2) to the ordinary differential equation derived from the finite difference method[8], the numerical results are also satisfactory.

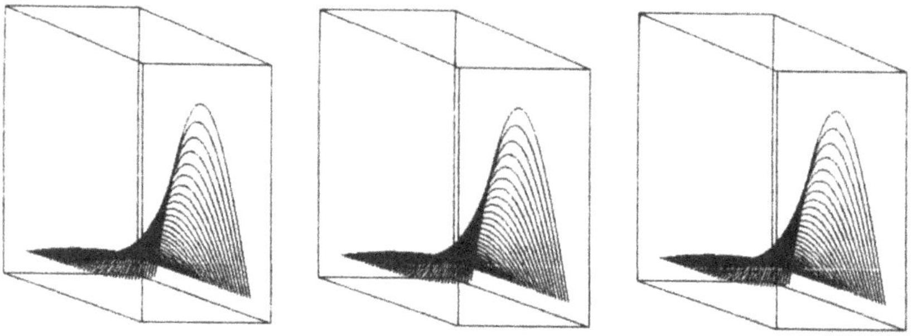

Crank Nicolson Scheme   Scheme (1)   Exact Solution

Figure 2

REFERENCES

1. Feng Kang, Formal Power Series and Numerical Algorithms for Dynamical Systems, "*Proc. Intern. Conf. on Scientific Computation, Hangzhou, 1991*", Ed. T. Chan, Z.C. Shi, (World Scientific, Singapore 1992), 28–35.
2. Haruo Yoshida, Construction of Higher Order symplectic Integrators, *Phys. Letters A* **150**:12(1990), 262–268
3. V.I.Arnold, Mathematical Methods of Classical Machanics. (Spinger-Verlag, 1978), 208–212.
4. Qin Meng-Zhao and Zhu Wen-jie, Construction of Higher Order Symplectic Schemes by Composition, *Computing* **47**(1992), 309–321.
5. Stanly Steinberg, Lie Series and Nonlinear Ordinary Differential Equations. *Journal of Mathematical Analysis And Applications*, Volume **101**, Number 1(1984), 39–63.
6. D. Gottlieb and S. Orszag, Numerical Analysis of Spectral Methods; Theory and Applications, *CBMS Regional Conference Series in Applied Mathematics* **26**(1977), (Society for Industrial and Applied Mathematics, Philadelphia).
7. Qin Meng-Zhao and Zhu Wen-jie, A Note on Stability of A Three-stage Difference Scheme for Ordinary Differential Equations, *Computers Math. Applic.* , Vol. **25**, No, 1(1993), 35–34.
8. Qin Meng-Zhao and Zhu Wen-jie, Applications of Higher Order Self-Adjoint Schemes to Partial Differential Equations(1992), preprint.
9. M. Suzuki, General Theory of Higher-Order Decomposition of Exponential Operators And Symplectic Integrators. *Phys. Letters A*, **165**(1992), 387–395.

NONLINEAR METHOD OF ORDINARY DIFFERENTIAL EQUATIONS AND ITS APPLICATIONS

QIN ZENG-FU

Mathematics Department, Fudan University
Shanghai, 200433, China

Abstract

Nonlinear method based on Frenet frame for ordinary differential equations is constructed. It applies to polycondensation kinetics equations, steady Burgers' equation and finding zeros of polynomial by homotopy.

1. Nonlinear method of ODEs

For IVP of ODE system

$$\begin{cases} \frac{dy}{dx} = f(x,y) \\ y(x^{(0)}) = y^{(0)}, \end{cases} \tag{1.1}$$

where $x \in R, y \in R^m$, the solution is a curve $y = y(x)$ in R^{m+1}. Set the additional dependent variable and corresponding right-hand function

$$y_0 = x, f_0(y_0, y_1, \cdots, y_m) \equiv 1, \tag{1.2}$$

and denote

$$Y = (y_0, y_1, \cdots, y_m)^T, F = (f_0, f_1, \cdots, y_m)^T. \tag{1.3}$$

Then, (1.1) is converted to the standard form

$$\begin{cases} \frac{dY}{dx} = F(Y) \\ Y(x^{(0)}) = Y^{(0)}. \end{cases} \tag{1.4}$$

Let s be the arclength of solution curve. According to Frenet equation, the unit tangent vector

$$e_1 = \dot{Y} = \frac{dY}{dx}\frac{dx}{ds} = \frac{1}{\lambda}F, \text{ where } \lambda = \|F\|; \tag{1.5}$$

the unit principal normal vector satisfies

$$\mathcal{K}e_2 = \dot{e}_1 = \frac{1}{\lambda^2}(U - \frac{q}{\lambda^2}F), \tag{1.6}$$

where

$$u_i = \sum_{j=0}^{m} \frac{\partial f_i}{\partial y_j} f_j, U = (u_0, u_1, \cdots, u_m)^T, \quad p = \|U\|, q = < F, U >, \tag{1.7}$$

and the curvature

$$\mathcal{K} = (\lambda^2 p^2 - q^2)^{\frac{1}{2}} / \lambda^3. \tag{1.8}$$

By further calculation [1], we get the following normal representation of the curve:

$$Y(s+\Delta s) = Y(s) + [\Delta s - \frac{\Delta s^3}{6}\mathcal{K}^2 - \frac{\Delta s^4}{8}\mathcal{K}\dot{\mathcal{K}}]e_1 + [\frac{\Delta s^2}{2}\mathcal{K} + \frac{\Delta s^3}{6}\dot{\mathcal{K}}$$
$$+ \frac{\Delta s^4}{24}(\ddot{\mathcal{K}} - \mathcal{K}^3 - \mathcal{K}\tau^2)]e_2 + [\frac{\Delta s^3}{6}\mathcal{K}\tau + \frac{\Delta s^4}{24}(2\dot{\mathcal{K}}\tau + \mathcal{K}\dot{\tau})]e_3$$
$$+ \frac{\Delta s^4}{24}\mathcal{K}\tau\sigma e_4 + O(\Delta s^5). \tag{1.9}$$

On this basis, the nonlinear method is constructed. As "go along the tangent and correct along the normal", we have the second order scheme

$$\tilde{Y} = Y + \frac{\Delta s}{\lambda}F + \frac{\Delta s^2}{2\lambda^2}(U - \frac{q}{\lambda^2}F). \tag{1.10}$$

Applying the Runge-Kutta idea, we get the two-stage second order scheme

$$\begin{bmatrix} \tilde{y}_0 \\ \tilde{y}_1 \\ \cdots \\ \tilde{y}_m \end{bmatrix} = \begin{bmatrix} y_0 \\ y_1 \\ \cdots \\ y_m \end{bmatrix} + \frac{\Delta s}{2}\left\{\frac{1}{\lambda}\begin{bmatrix} f_0 \\ f_1 \\ \cdots \\ f_m \end{bmatrix} + \frac{1}{\lambda^*}\begin{bmatrix} f_0^* \\ f_1^* \\ \cdots \\ f_m^* \end{bmatrix}\right\}, \tag{1.11}$$

where

$$f_j^* = f_j(y_0 + \frac{\Delta s}{\lambda}f_0, y_1 + \frac{\Delta s}{\lambda}f_1, \cdots, y_m + \frac{\Delta s}{\lambda}f_m), \lambda^* = \|F^*\|. \tag{1.12}$$

There are a family of the fourth order schemes. Specially, we propose the following fourth order scheme

$$\begin{bmatrix} \tilde{y}_0 \\ \tilde{y}_1 \\ \cdots \\ \tilde{y}_m \end{bmatrix} = \begin{bmatrix} y_0 \\ y_1 \\ \cdots \\ y_m \end{bmatrix} + \frac{\Delta s}{\lambda}\begin{bmatrix} f_0 \\ f_1 \\ \cdots \\ f_m \end{bmatrix} + \frac{\Delta s^2}{6}\left\{[\frac{1}{\lambda^2}\begin{bmatrix} u_0 \\ u_1 \\ \cdots \\ u_m \end{bmatrix} - \frac{q}{\lambda^4}\begin{bmatrix} f_0 \\ f_1 \\ \cdots \\ f_m \end{bmatrix}]\right.$$
$$\left. + 2[\frac{1}{\hat{\lambda}^2}\begin{bmatrix} \hat{u}_0 \\ \hat{u}_1 \\ \cdots \\ \hat{u}_m \end{bmatrix} - \frac{\hat{q}}{\hat{\lambda}^4}\begin{bmatrix} \hat{f}_0 \\ \hat{f}_1 \\ \cdots \\ \hat{f}_m \end{bmatrix}]\right\}, \tag{1.13}$$

where $\hat{f}_j$ and $\hat{u}_j$ are evaluated at the point

$$(y_0 + \frac{\Delta s}{\lambda}f_0 + \frac{\Delta s^2}{8}(\frac{u_0}{\lambda^2} - \frac{q}{\lambda^4}f_0), \cdots, (y_m + \frac{\Delta s}{\lambda}f_m + \frac{\Delta s^2}{8}(\frac{u_m}{\lambda^2} - \frac{q}{\lambda^4}f_m))$$

and

$$\hat{l} = \|\hat{F}\|, \qquad \hat{q} = <\hat{F}, \hat{U}> . \tag{1.14}$$

To ensure the absolute stability, the permissible stepsize is

$$\Delta s_{perm} = \frac{4(\lambda^2 - 1)}{\mathcal{K}\lambda^2(\lambda^2 + 1)}. \tag{1.15}$$

This stepsize criterion reflects the geometric nature of the solution curve: the large curvature $\mathcal{K}$ indicates the sharp turn of this curve, so the stepsize must be small; the large steepness λ indicates the remarkable drop of this curve, so the stepsize should be small enough.

2. Polycondensation kinetics equations

The polycondensation kinetics can be reduced to the IVP of ODEs

$$\begin{cases} \dfrac{dz}{dx} = -R_1 z^2 - R_2 z(y-z) \\ \dfrac{dy}{dx} = -\frac{1}{2}R_1 z^2 - R_2 z(y-z) - \frac{1}{2}(y-z)^2 \\ z(0) = y(0) = 1, \end{cases} \tag{2.1}$$

where z and y is the relative concentration of monomer and polymers altogether; R_1 and R_2 is the relative rate constant. For the polycondensation kinetics of polyphenylene ether sulphone [2] at $120°C$, $R_1 = 12.0$ and $R_2 = 0.66$.

Corresponding to the stiff ratio, we take the boundary layer width $\tau = 1.0$ and the observation interval length $T = 5.5$. In the boundary layer $[0, \tau]$, the numerical solution is obtained by the fourth order nonlinear method accurately. Out of the boundary layer, due to the decay of the derivative

$$\left|\frac{dz}{dx}\right|_{x \geq \tau} \leq \frac{1}{N} \max_{[0,\tau]} \left|\frac{dz}{dx}\right|, \text{ where } N = 100 \tag{2.2}$$

we make use of the fitting

$$\frac{dz}{dx} \approx -Qz^2, \qquad x \in [\tau, T] \tag{2.3}$$

to get the algebraic-differential equation system

$$\begin{cases} -Qz^2 = -R_1 z^2 - R_2(y-z)z \\ \dfrac{dy}{dx} = -\frac{1}{2}R_1 z^2 - R_2(y-z)z - \frac{1}{2}(y-z)^2. \end{cases} \qquad x \in [\tau, T] \tag{2.4}$$

The above fitting coincides with the behavior $\dfrac{dz}{dx} < 0$, $\dfrac{d^2 z}{dx^2} > 0$ and $\dfrac{d^3 z}{dx^3} < 0$. And the resulting algebraic-differential equation system can be solved by the same fourth order nonlinear method with greater stepsize.

3. Steady Burgers' equation

The steady Burgers' equation

$$u\frac{du}{dx} = \frac{1}{\text{Re}}\frac{d^2u}{dx^2} \tag{3.1}$$

express the balance between convection term $u\dfrac{du}{dx}$ and diffusion term $\dfrac{1}{\text{Re}}\dfrac{d^2u}{dx^2}$ in fluid dynamics, where Re is the Reynold's number. This ODE has the representative solution

$$u = -\text{th}(\frac{\text{Re}}{2}x). \tag{3.2}$$

In this case, there is a shock wave layer about $x = 0$.

As twopoint BVP

$$\begin{cases} u\dfrac{du}{dx} = \dfrac{1}{\text{Re}}\dfrac{d^2u}{dx^2} \\ u(-1) = -\text{th}(-\frac{\text{Re}}{2}), u(1) = -\text{th}(\frac{\text{Re}}{2}), \end{cases} \tag{3.3}$$

we make use of shooting. Set up corresponding IVP

$$\begin{cases} \dfrac{du}{dx} = v \\ \dfrac{dv}{dx} = \text{Re } uv \\ u(-1) = -\text{th}(-\dfrac{\text{Re}}{2}), v(-1) = \eta, \end{cases} \tag{3.4}$$

where η ia a parameter. For the fictitious funtion

$$f(\eta) = u(\eta, 1) - u(1), \tag{3.5}$$

we are to find η^* such that

$$f(\eta^*) = 0. \tag{3.6}$$

For this end we construct the following homotopy [3]:

$$H(\lambda, \eta) = (1-\lambda)\eta + \lambda f(\eta), \qquad \lambda \in [0,1]. \tag{3.7}$$

Along the zero curve $H(\lambda, \eta) = 0$, the following IVP holds

$$\begin{cases} \dfrac{d\lambda}{d\sigma} = \dfrac{\alpha}{\gamma} \\ \dfrac{d\eta}{d\sigma} = \dfrac{\beta}{\gamma} \\ \lambda(0) = \eta(0) = 0, \end{cases} \tag{3.8}$$

where σ is the arclength of zero curve, and

$$\alpha = (1-\lambda) + \lambda D_\eta f(\eta), \beta = \eta - f(\eta), \gamma = (\alpha^2 + \beta^2)^{\frac{1}{2}}, \tag{3.9}$$

and $D_\eta f(\eta) = D_\eta u(\eta, 1)$ can be obtained from the system

$$\begin{cases} \dfrac{d}{dx}(D_\eta u) = D_\eta v \\ \dfrac{d}{dx}(D_\eta v) = \text{Re}(v D_\eta u + u D_\eta v) \\ D_\eta u(\eta, -1) = 0, D_\eta v(\eta, -1) = 1. \end{cases} \tag{3.10}$$

In order to guarantee the accuracy of the numerical solution in shock wave layer, we use the fourth order nonlinear method to compute the IVP(3.4). In the homotopy algorithm we use the second order nonlinear method to solve the system (3.8) and (3.10) synchronously.

In the case Re = 10, the corresponding numerical results [4]

$$\tilde{\eta} = -0.0009075, \quad u(\tilde{\eta}, 1) = -0.99990912$$

approximate their exact values respectively

$$\eta^* = -0.0009079162, \quad u(1) = -0.9999092043.$$

4. Zeros of polynomial

For finding the zeros of the polynomial

$$p(z) = a_0 z^n + a_1 z^{n-1} + \cdots + a_{n-1} z + a_n. \tag{4.1}$$

we construct the following homotopy

$$H(z, w, t) = (1-t)(z^n - w) + tP(z), t \in [0,1], \tag{4.2}$$

where w is some complex number

$$w = a + ib. \tag{4.3}$$

Obviously,

$$H(z, w, 0) = 0 \Longleftrightarrow z^n - w = 0, H(z, w, 1) = 0 \Longleftrightarrow p(z) = 0.$$

Denote

$$z = x + iy, \tag{4.4}$$

the real form of homotopy is

$$\begin{aligned} &H_1(x, y, a, b, t) = \mathrm{Re}\{H(z, w, t)\} \\ =&(1-t)\left\{\sum_{j=0}^{[\frac{n}{2}]} C_n^{2j}(-1)^j y^{2j} x^{n-2j} - a\right\} + t\sum_{k=0}^{n} a_{n-k} \sum_{j=0}^{[\frac{k}{2}]} C_k^{2j}(-1)^j y^{2j} x^{k-2j}, \\ &H_2(x, y, a, b, t) = \mathrm{Im}\{H(z, w, t)\} \\ =&(1-t)\left\{\sum_{j=0}^{n-1-[\frac{n}{2}]} C_n^{2j+1}(-1)^j y^{2j+1} x^{n-(2j+1)} - b\right\} \\ &+ t\sum_{k=0}^{n} a_{n-k} \sum_{j=0}^{k-1-[\frac{k}{2}]} C_k^{2j+1}(-1)^j y^{2j+1} x^{k-(2j+1)}. \end{aligned} \tag{4.5}$$

Set

$$\alpha = \frac{\partial(H_1, H_2)}{\partial(y, t)}, \beta = \frac{\partial(H_1, H_2)}{\partial(t, x)}, \gamma = \frac{\partial(H_1, H_2)}{\partial(x, y)}, \quad \delta = (\alpha^2 + \beta^2 + \gamma^2)^{\frac{1}{2}}. \tag{4.6}$$

According to Sard theorem [3], for almost all w, we have $\gamma > 0$. Along the zero curve

$$\begin{cases} H_1(x, y, a, b, t) = 0 \\ H_2(x, y, a, b, t) = 0, \end{cases} \tag{4.7}$$

taking the arclength σ as independent variable, we get the differential equations and corresponding initial values

$$\begin{cases} \dfrac{dx}{d\sigma} = \dfrac{\alpha}{\delta}, \dfrac{dy}{d\sigma} = \dfrac{\beta}{\delta}, \dfrac{dt}{d\sigma} = \dfrac{\gamma}{\delta}, \\ x(0) = \mathrm{Re}\{\sqrt[n]{w}\}, y(0) = \mathrm{Im}\{\sqrt[n]{w}\}, \quad t(0) = 0. \end{cases} \tag{4.8}$$

when $t = 1$, the corresponding $x + iy$ are zeros of the polynomial. The above IVP of ODEs is solved successfully by the fourth order nonlinear method.

As an example, the eigenvalue problem of matrix

$$A = \begin{bmatrix} -5.509882 & 1.870086 & 0.422908 & 0.008814 \\ 0.287865 & -11.811654 & 5.711900 & 0.058717 \\ 0.049099 & 4.308033 & -12.970687 & 0.229326 \\ 0.006235 & 0.269851 & 1.397369 & -17.596207 \end{bmatrix} \tag{4.9}$$

leads to its characteristic polynomial

$$p(z) = \det(zI - A) = z^4 + 47.88843z^3 + 797.2789z^2 + 5349.457z + 12296.555. \tag{4.10}$$

The numerical results given by the fourth order nonlinear method

$$\tilde{z}_1 = -17.8632, \tilde{z}_2 = -17.1524, \tilde{z}_3 = -7.5741, \tilde{z}_4 = -5.2987$$

approximate four zeros

$$z_1 = -17.863034, z_2 = -17.15266, z_3 = -7.574045, z_4 = -5.29870.$$

References

[1] Qin Zeng-fu, A class of nonlinear methods for ordinary differential equations, JCM, 3(1985), 320-327.

[2] Yu Tong-yin et al., Polycondensation kinetics of polyphenylene ether sulphone, Polymer, 25(1984), 1363-1366.

[3] E. L. Allgower, A survey of homotopy methods for smooth mappings, Lecture Notes in Math. 878, 1981, 1-29.

[4] Qin Zeng-fu, On the numerical solution of steady Burgers' equation, Proceedings of BAIL-V, 1988, 295-298.

Decomposition of Vector Fields and Composition of Algorithms*

Wang Daoliu
Computing Center, Chinese Academy of Sciences,
Beijing 100080, China

Abstract
In this paper we give a method to construct explicit structure-preserving algorithms for a kind of so called separable systems with some Lie algebraic structure.

1. Introduction

All vector fields on $\mathbf{R}^m$ form a Lie algebra $\mathbf{V}(m)$ under the Lie bracket $[a,b] = a_*b - b_*a$, where a_* is the Jacobian matrix of the vector field a. The associated Lie group $\mathbf{D}(m)$ consists of diffeomorphisms on $\mathbf{R}^m$. A vector field a defines a dynamical system (or the system of ODEs)

$$\frac{dx}{dt} = a(x), \qquad x \in \mathbf{R}^m. \tag{1}$$

If a belongs to a Lie subalgebra $\mathcal{L} \subset \mathbf{V}$, then the phase flow of (1) is a one parameter group of the associated Lie subgroup $\mathcal{G} \subset \mathbf{D}$. An algorithm to solve System (1) determines a step transition operator g_a^s, where s is the step-size.

Definition 1 (**[1,2]**) *Let $\mathcal{L}$ be a Lie subalgebra of $\mathbf{V}$ and $\mathcal{G}$ the associated Lie subgroup of $\mathbf{D}$. A vector field $a \in \mathcal{L}$. If $g_a^s \in \mathcal{G}$, then g_a^s, or the corresponding algorithm, is called structure-preserving.*

It is natural to require the numerical algorithms solving (1) to be structure-preserving. The typical examples of structure-preserving algorithms are symplectic algorithms for Hamiltonian systems. The volume-preserving algorithms for source-free systems and contact algorithms for contact systems have been constructed in [4,5]. In this paper, we generalize the method of decomposition of vector fields used in [3] for a kind of Hamiltonian systems to general systems with some Lie algebraic structures.

*Supported by National Natural Science Foundation of China and State Commission of Science and Technology.

2. Decomposition of vector fields and composition of algorithms

Definition 2 *A vector field a is nilpotent of degree 2 if $a_* a = 0$.*

Lemma 1 *If a is nilpotent of degree 2, then the phase flow of the system (1) is*

$$e_a^t(x) = x + ta(x). \tag{2}$$

Lemma 2 *For a nilpotent vector field a, the explicit Euler method is structure-preserving.*

Lemma 1 can be easily verified directly. Lemma 2 follows from Lemma 1 since the right hand side of Eq. (2) is just the explicit Euler method.

Definition 3 *Let $\mathcal{L}$ be a Lie subalgebra of vector fields. A vector field $a \in \mathcal{L}$ is $\mathcal{L}$-separable if a has the following decomposition*

$$a = a_1 + a_2 + \cdots + a_k, \qquad a_i \in \mathcal{L}, \quad (a_i)_* a_i = 0, \quad i = 1, \cdots, k. \tag{3}$$

That means, if $a \in \mathcal{L}$ can be decomposed as the sum of nilpotent vector fields with degree 2 in $\mathcal{L}$, then a is $\mathcal{L}$-separable.

Theorem 1 *If a is $\mathcal{L}$-separable and admits the decomposition (3), then the algorithm*

$$g_a^s = E_{a_{i_1}}^s \circ E_{a_{i_2}}^s \circ \cdots \circ E_{a_{i_k}}^s \tag{4}$$

is an explicit structure-preserving algorithm of order 1, where $(i_1, i_2, \cdots, i_k)$ is any permutation of $(1, 2, \cdots, k)$.

Proof. For the sake of simplicity, we assume $(i_1, i_2, \cdots, i_k) = (k, k-1, \cdots, 2, 1)$. Since $E_{a_i}^s$ is the phase flow of the system corresponding to the vector field a_i, $E_{a_i}^s \in \mathcal{G}$, for all s. Therefore, $\prod_{i=1}^{k} E_{a_i}^s \in \mathcal{G}$. That is, g_a^s is structure-preserving. Expanding g^s with respect to s, we obtain

$$\begin{aligned} g_a^s &= (1 + sa_k + O(s^2)) \circ \cdots \circ (1 + sa_1 + O(s^2)) \\ &= 1 + s(a_1 + a_2 + \cdots + a_k) + O(s^2) = 1 + sa + O(s^2). \end{aligned}$$

This implies that g^s is of order 1. ∎

Algorithm (4): $x \to \hat{x} = g_a^s(x)$ is performed by the following procedure: For given $x = x^0$, $x^j = E_{a_j}^s x^{j-1} = x^{j-1} + sa_j(x^{j-1})$, for $j = 1, \cdots, k$. Then $\hat{x} = x^k$ gives the new one.

In order to construct higher order structure-preserving algorithms, we recall the concept of revertibility of algorithms. The detailed materials can be referred to, such as, [1,3,8,9]. An algorithm is revertible if

$$g^{-s} \circ g^{s} = 1.$$

For any g^s, $(g^{-s})^{-1} \circ g^s$ and $g^s \circ (g^{-s})^{-1}$ are always revertible. Revertible g^s is always of even order of accuracy when it approximates e^s, i.e.,

$$g^{s} = e^{s} + O(s^{2r+1}), \qquad r \geq 0.$$

Lemma 3 ([7,9,6]) *If g^s is revertible and of order 2, then $g^{\alpha s} \circ g^{\beta s} \circ g^{\alpha s}$ is of order 4 when $\alpha = (2-2^{1/3})^{-1}$, $\beta = -2^{1/3}(2-2^{1/3})^{-1}$.*

Lemma 4 ([9,8]) *If g^s is revertible and of order $2r$, then $g^{\alpha s} \circ g^{\beta s} \circ g^{\alpha s}$ is of order $2(r+1)$ when*

$$2\alpha + \beta = 1, \qquad 2\alpha^{2r+1} + \beta^{2r+1} = 0,$$

i.e.,

$$\alpha = \frac{1}{2 - 2^{1/(2r+1)}}, \qquad \beta = -\frac{2^{1/(2r+1)}}{2 - 2^{1/(2r+1)}}. \tag{5}$$

It follows that for $\mathcal{L}$-separable vector field a of the form (3)

$$g_2^s = (g_a^{-s/2})^{-1} \circ g_a^{s/2} = E_{a_1}^{s/2} \circ \cdots \circ E_{a_{k-1}}^{s/2} \circ E_{a_k}^{s} \circ E_{a_{k-1}}^{s/2} \circ \cdots \circ E_{a_1}^{s/2} \tag{6}$$

is an explicit, structure-preserving algorithm of order 2. And

$$g_4^s = g_2^{\alpha s} \circ g_2^{\beta s} \circ g_2^{\alpha s}, \qquad \alpha = (2-2^{1/3})^{-1}, \quad \beta = -2^{1/3}(2-2^{1/3})^{-1} \tag{7}$$

is an explicit structure-preserving algorithm of order 4.

Remark: In fact, if $a \in \mathcal{L}$ can be decomposed as

$$a = a_1 + a_2 + \cdots + a_k, \qquad a_i \in \mathcal{L}, \quad i = 1, \cdots, k, \tag{8}$$

and $e_{a_i}^t$ are algorithmically performable, then

$$g_a^s = e_{a_k}^s \circ \cdots \circ e_{a_1}^s \tag{9}$$

is also a structure-preserving algorithm of order 1.

3. Applications

The method mentioned above is very simple but it can cover many important cases.

3.1. Hamiltonian systems

If a is a Hamiltonian vector field on $\mathbf{R}^{2n}$, i.e.,

$$a(z) = J\nabla H(z), \qquad J = \begin{bmatrix} 0 & -1 \\ 1 & 0 \end{bmatrix}, \qquad z \in \mathbf{R}^{2n}$$

for some Hamiltonian function H, where ∇H is the gradient of H, $\nabla H = (H_{z_1}, H_{z_2}, \cdots, H_{z_{2n}})^T$. Then

$$a_* a = JH_{zz}J\nabla H, \qquad H_{zz}, \quad \text{the Hessian of } H.$$

a is nilpotent of degree 2 if and only if

$$H_{zz}J\nabla H = 0. \tag{10}$$

Therefore, a Hamiltonian function H is nilpotent of degree 2 if H satisfies (10). A Hamiltonian vector field a is separable in Hamiltonian Lie algebra, or in short, H-separable, if and only if there are Hamiltonians H_i, $i = 1, \cdots, k$ such that

$$a = \sum_{i=1}^{k} a_i = \sum_{i=1}^{k} J\nabla H_i, \quad J(H_i)_{zz}J\nabla H_i = 0.$$

Consequently, a Hamiltonian H is H-separable if and only if

$$H = \sum_{i=1}^{k} H_i, \qquad (H_i)_{zz}J\nabla H_i = 0, \quad i = 1, \cdots, k.$$

This is just the case discussed in [1,3]. For more details about separable Hamiltonian systems and their structure-preserving, i.e., symplectic, algorithms, refer to [1,3].

3.2. Source-free systems

A system

$$\frac{dx}{dt} = a(x) \tag{11}$$

is source-free if $\operatorname{div} a = 0$. The set of all source-free vector fields forms a Lie subalgebra. The associated Lie subgroup is the set of volume-preserving mappings. The phase flow of a source-free system is a one-parameter group of volume-preserving mappings. Hence, for source-free systems, structure-preserving algorithms are just the volume-preserving algorithms. Usual conventional algorithms are not volume-preserving even for linear source-free systems. Preliminary versions of volume-preserving algorithms for general source-free systems have been given in [4]. Here we give a special version for a kind of separable source-free systems.

Let $a(x) = (a_1(x), \cdots, a_m(x))^T$ be a source-free vector field. If $\frac{da_i}{dx_i} = 0$, i.e., each component $a_i(x)$ is independent of x_i, then $a^{(i)} =: (0, \cdots, 0, a_i, 0, \cdots, 0)^T$, i.e., only the i-th component is a_i, others are zero, is source-free. It is also nilpotent of degree 2, i.e.,

$$a_*^{(i)} a^{(i)} = 0.$$

Hence, such a vector field a is separable,

$$a = a^{(1)} + a^{(2)} + \cdots + a^{(m)}. \tag{12}$$

The explicit scheme of order 1

$$g_1^s = E^s_{a^{(m)}} \circ E^s_{a^{(m-1)}} \circ \cdots \circ E^s_{a^{(1)}} \tag{13}$$

is volume-preserving. The permutations of $a^{(i)}$ lead to other explicit volume-preserving schemes of order 1. The scheme

$$g_2^s = E^{s/2}_{a^{(1)}} \circ \cdots \circ E^{s/2}_{a^{(m-1)}} \circ E^s_{a^{(m)}} \circ E^{s/2}_{a^{(m-1)}} \circ \cdots \circ E^{s/2}_{a^{(1)}} \tag{14}$$

is revertible, therefore is an explicit volume-preserving scheme of order 2. Explicit volume-preserving schemes with higher order can be given by Lemmas 3 and 4.

Scheme (13) is performed by the following procedure.

Let $x^k = (x_1^k, \cdots, x_m^k)$ be given. Then

$$\begin{aligned}
x_1^{k+1} &= x_1^k + s a_1(x_2^k, x_3^k, \cdots, x_m^k), \\
x_2^{k+1} &= x_2^k + s a_2(x_1^{k+1}, x_3^k, \cdots, x_m^k), \\
&\cdots \\
x_m^{k+1} &= x_m^k + s a_m(x_1^{k+1}, x_2^{k+1}, \cdots, x_{m-1}^{k+1}).
\end{aligned}$$

It is like the Jacobian iteration. Similarly, Scheme (14) is performed by: If $x^k = (x_1^k, \cdots, x_m^k)$ is given then

$$\begin{aligned}
\hat{x}_1 &= x_1^k + \frac{s}{2} a_1(x_2^k, x_3^k, \cdots, x_m^k), \\
\hat{x}_2 &= x_2^k + \frac{s}{2} a_2(\hat{x}_1, x_3^k, \cdots, x_m^k), \\
&\cdots \\
\hat{x}_{m-1} &= x_{m-1}^k + \frac{s}{2} a_{m-1}(\hat{x}_1, \cdots, \hat{x}_{m-2}, x_m^k), \\
x_m^{k+1} &= x_m^k + s a_m(\hat{x}_1, \hat{x}_2, \cdots, \hat{x}_{m-1}), \\
x_{m-1}^{k+1} &= \hat{x}_{m-1} + \frac{s}{2} a_{m-1}(\hat{x}_1, \hat{x}_2, \cdots, x_m^{k+1}), \\
&\cdots \\
x_1^{k+1} &= \hat{x}_1 + \frac{s}{2} a_1(x_1^{k+1}, x_2^{k+1}, \cdots, x_{m-1}^{k+1}).
\end{aligned}$$

Let

$$\frac{dx}{dt} = Ax, \qquad \operatorname{tr} A = \sum a_{ii} = 0 \tag{15}$$

be a linear source-free vector field. Decompose A as

$$A = D + L + U,$$

where L and U are strictly lower and upper triangular matrices respectively, $D = \operatorname{diag}(a_{11}, \cdots, a_{mm})$. Therefore they are also source-free. If $D \neq 0$ then A is not separable in the above sense but $L + U$ is. Scheme (13) of the matrix form,

$$g^s_{L+U} = (I - sL)^{-1}(I + sU),$$

is a volume-preserving algorithm of order 1 for the system determined by $L+U$. D is diagonal. Its phase flow is e^{tD}. Therefore

$$\begin{aligned} g^s_{A,1} &= e^{sD}(I - sL)^{-1}(I + sU), \\ g^s_{A,2} &= \left(I - \frac{s}{2}U\right)^{-1}\left(I + \frac{s}{2}L\right)e^{sD}\left(I - \frac{s}{2}L\right)^{-1}\left(I + \frac{s}{2}U\right) \end{aligned}$$

are volume-preserving algorithms of order 1 and 2 respectively. Higher order volume-preserving algorithms can be obtained by Lemmas 3 and 4.

References

[1] Feng Kang, Wang Daoliu, Dynamical Systems and Algorithms for Ordinary Differential Equations, Preprint, Acad. Sin. Comp. Ctr., 1991.

[2] Feng Kang, Symplectic, contact and volume-preserving algorithms, *Proc. 1st China-Japan Conference on Numerical Mathematics*, Beijing, 1992, World Scientific, Singapore, 1993.

[3] Feng Kang, Wang Daoliu, Variations on a theme by Euler, Preprint, Acad. Sin. Comp. Ctr., 1991.

[4] Feng Kang, Shang Zaijiu, Volume preserving algorithms for systems with divergence-free vector fields, Preprint, A.S.C.C., 1991.

[5] Feng Kang, Contact algorithms for contact systems, Preprint, A.S.C.C., 1991.

[6] Qin Meng-zhao, Wang Dao-liu and Zhang Mei-qing, Explicit symplectic difference schemes for separable Hamiltonian systems, *J. Comp. Math.*, **9**:**3**(1991), 211–221.

[7] Qin Meng-zhao, Zhang Mei-qing, Multi-stage symplectic schemes of two kinds of Hamiltonian systems for wave equation, *Computers Math. Applic.*, **19**:10 (1990), 51–52.

[8] Qin Meng-zhao, Zhu Wen-jie, Construction of higher order schemes by composition, *Computing*, **47**(1992), 309–321.

[9] H. Yoshida, Construction of higher order symplectic integrators, *Phys. Letters A*, **150**(1990), 262–268.

UNIFIED MULTI-DIMENSIONAL SIMULATION TECHNIQUE FOR SOLID, LIQUID AND GAS

TAKASHI YABE
Gunma University, Tenjin-chou, Kiryu, Gunma 376, Japan

ABSTRACT

A new higher-order hyperbolic scheme is proposed. This scheme is applicable simultaneously to compressible and incompressible fluids. A sharp interface of two fluids is described by one grid throughout computation. The code is also applied to laser-induced evaporation dynamics.

1. Introduction

The interaction of compressible gas with incompressible liquid or solid is of primary interest from the view point both of numerical technique and physics. The numerical schemes for two different phases have independently been developed. For compressible fluid, elaborate schemes like TVD or ENO proved to be quite effective in capturing shock waves. However, since these schemes employ conservative form of fluid equation, divergence of velocity which becomes zero in incompressible limit cannot be treated independently of the advection part. On the contrary, incompressible schemes like QUICK or higher-order upwind schemes combined with improved MAC (Marker and Cell) procedure can treat divergence-free fluid vorticity and turbulence. However, these schemes cannot always treat a shock wave with a sharp discontinuity. We need a scheme to treat both compressible and incompressible fluids simultaneously in one program to simulate the interaction of gas with liquid. For this purpose, we must treat compressible fluid by nonconservative equations and treat incompressible fluid by the same equation without imposing any condition like divergence free. Fully implicit solver can treat this procedure, but the convergence of iteration in highly distorted state is still a problem.

Recently, we have proposed a new type of scheme (CIP)[1,2,3] to treat the shock wave by a non-conservative scheme. A similar effort has been done by Wu with SOMS scheme[4]. Since the CIP does not tranform physical quantities into characteristic invariants, it is quite easily extended to variety of equations. By a simple extention, the CIP can be used both for compressible and incompressbile fluids. This code is called CCUP[5] and can treat incompressible fluid with full Euler equations.

We report here two examples to demonstrate the ability of this scheme. One is the injection of fluid into another fluid. In this case, a sharp interface is described by a density function and two fluids are separated by one grid. Another example is the laser-induced evaporation dynamics. A metal is irradiated by a laser and is melted and evaporated.

2. Interface Tracking of Two Fluids

We use the full Euler equations both for compressible and incompressible fluids.

$$\frac{\partial \rho}{\partial t} + (\vec{u} \cdot \nabla)\rho = -\rho \nabla \cdot \vec{u}, \tag{1}$$

$$\frac{\partial \vec{u}}{\partial t} + (\vec{u} \cdot \nabla)\vec{u} = -\frac{\nabla P}{\rho} + Q_u, \tag{2}$$

$$\rho[\frac{\partial E}{\partial t} + (\vec{u} \cdot \nabla)E] = -P\nabla \cdot \vec{u} + Q_E, \tag{3}$$

where ρ is the density, $\vec{u}$ the velocity, P the pressure, E the specific internal energy, Q_u represents viscous and stress terms, and Q_E includes heat conduction and heat source.

There are two independent treatments of interaction of compressible fluid with incompressible fluid. One is to separate two fluids by an interface. This problem is treated in this section. In another case, one fluid undergoes phase transition into another fluid. One example is the evaporation process. In this case, two fluids can penetrate each other and hence we need an equation of state to describe phase change. This case will be treated in the next section.

There have been numerous methods proposed for treating interface between two different materials. These methods are divided into two groups. In one group the interface is described by a surface function, while in the other group the interface is defined as surface of a density function[6]. In the former case, main problem arises from a multi-valued function when the surface is strongly distorted or even breaks up. Although this shortcoming will not arise in the latter case, the numerical scheme to describe an evolution of the density function without numerical diffusion is a problem which needs further investigation.

In this paper, we propose a novel method to treat this density function with high accuracy in multi-dimensions. For this purpose, we use the CIP (Cubic-Interpolated Propagation) method recently developed by the author. The CIP method can trace a very sharp interface. However, the method is still not enough for the problem which we are going to present. The interface must be treated by one grid over the whole calculation. Therefore, we modify the CIP method slightly. By this modification, we have succeeded to trace a sharp interface by one grid throughout the calculation.

Here we use a density function ϕ to describe the abundance of a material A. Therefore, we define

$$\begin{aligned} \phi &= 1 \quad \text{(Material A)} \\ &= 0 \quad \text{(No material or different material)} \end{aligned} \tag{4}$$

This density fuction evolves according to

$$\frac{\partial \phi}{\partial t} + u_i \frac{\partial \phi}{\partial x_i} = 0, \tag{5}$$

where u_i is the fluid velocity component. In the CIP method, this function inside a grid cell is approximated by a cubic polynomial in three spatial directions. Furthermore, the spatial derivatives of Eq.(5)

$$\frac{\partial}{\partial t}(\frac{\partial \phi}{\partial x_j}) + u_i \frac{\partial}{\partial x_i}(\frac{\partial \phi}{\partial x_j}) = -(\frac{\partial \phi}{\partial x_i})(\frac{\partial u_i}{\partial x_j}), \tag{6}$$

are also used to determine the polynomial. That is the dependent variables are ϕ and $\partial\phi/\partial x_j$ and these are determined from Eqs.(5) and (6). Other unknown coefficients to determine this polynomial are found by imposing continuity of the value and the spatial derivatives among neighboring cells. We will not describe it in detail but the solution in one-, two- and three-dimensions are given in Ref.3.

Propagation of a linear wave has been described with high accuracy by this method. However, diffusion and undulation at a sharp front is unavoidable. The previous method to suppress those is successful only in one dimension but is complicated[2,3]. In the problem of interface tracking, ϕ is always limited in the range from 0 to 1 and hence we can use much more effective method to suppress them. We propose here to transform ϕ into $F(\phi)$. It is obvious that a new function $F(\phi)$ also obeys the same equation as Eq.(5).

$$\frac{\partial F}{\partial t} + u_i \frac{\partial F}{\partial x_i} = 0, \tag{7}$$

Therefore, all the method proposed for ϕ in the CIP method can also be used for this F. What is the merit of this transformation ? Let us propose a tangent function for F as

$$F = \tan[0.99\pi(\phi - 0.5)],$$

$$\phi = (\arctan F)/(0.99\pi) + 0.5. \tag{8}$$

Although F may be slightly diffusive and may have undulation when Eq.(7) is solved, ϕ is always limited between 0 and 1. Furthermore, most of values are concentrated near $\phi = 0$ and 1. Therefore, monotone and sharp discontinuity can be described quite easily. The factor 0.99 in this transformation is to avoid the inifinite value when $\phi = 0, 1$.

In order to show how it works, we show a result of a linear wave propagation. In Fig.1(b), we show a result of two dimensional rotation of solid body by the CIP0 method described in the previous paper[3]. Although this result is comparable with and even better than other methods, we can further improve this result by simply using a transformation proposed in Eq.(8). Figure 1(c) is the result of the present scheme. Surprisingly, the shape of the function is preserved throughout the computation.

This scheme is applied to the injection of heavier fluid into lighter fluid as was previously investigated by the author[7]. In the present case, we set the sound speed to infinite value and hence the process treated here is purely incompressible. The result is shown in Fig.2. The interface of two fluids has successfully been treated by one grid throughout the computation.

3. Laser-Induced Evaporation Dynamics

Now, we show another example with the CCUP[8]. In simulations, a semi-infinite aluminum slab is illuminated by a laser, whose pulse is triangular in time and peaks at 3 μsec. The focal spot diameter is 1 mm. Laser light is assumed to be deposited at the surface of metal. Absorbed laser intensity is 50 MW/cm^2.

Figure 3 shows the time sequence of the density contour thus obtained. In the course of time, the evaporated metal vapor is accumulated near the surface and may absorb the laser light. At the metal surface, we found that whisker-like structure appears. This may be the melting instability found by Anisimov[9]. Although Anisimov did not consider the effect of metal vapor in his theory, it can straightforwardly be extended to the present case. Let us assume that the laser energy is deposited at a point far from the metal surface. Because of thermal conduction, the temperature decreases from the point of absorption to the metal surface, whose temperature is assumed to be slightly below the melting temperature. When the surface moves slightly towards the laser beam, that is , towards the direction that the temperature increases, the surface should be heated more up to the melting temperature. Thus, the moving towards the high temperature regions is beneficial for evaporation. Figure 3 clearly shows that all clusters of melted materials from the surface are moving towards the point of laser absorption.

References

1. H.Takewaki, A.Nishiguchi and T.Yabe, *J.Comput. Phys.* 61(1985)261.
2. T.Yabe and E.Takei, *J.Phys.Soc.Japan* 57(1988)2598.
3. T.Yabe *et al.*, *Comput. Phys. Comm.* 66(1991)219 and 233.
4. H.M.Wu and M.L.Yang, *Lecture Note*,No.141 (1980) 442.
5. T.Yabe and P.Y.Wang, *J.Phys.Soc.Japan* 60 (1991)2105.
6. C.W.Hirt and B.D.Nichols, *J.Comput.Phys.* 39 (1981).
7. T.Yabe, H.Hoshino and T.Tsuchiya, *Phys. Rev.* **A44** (1991)2756.
8. T.Yabe, T.Mochizuki and H.Hara, *Proc. of International Conference Laser Advanced Materials Processing, June, Nagaoka*, (1992)387.
9. S.I.Anisimov, M.I.Tribel'skii and Ya.G.Epel'baum, *Sov. Phys. JETP* 51 (1980)802.

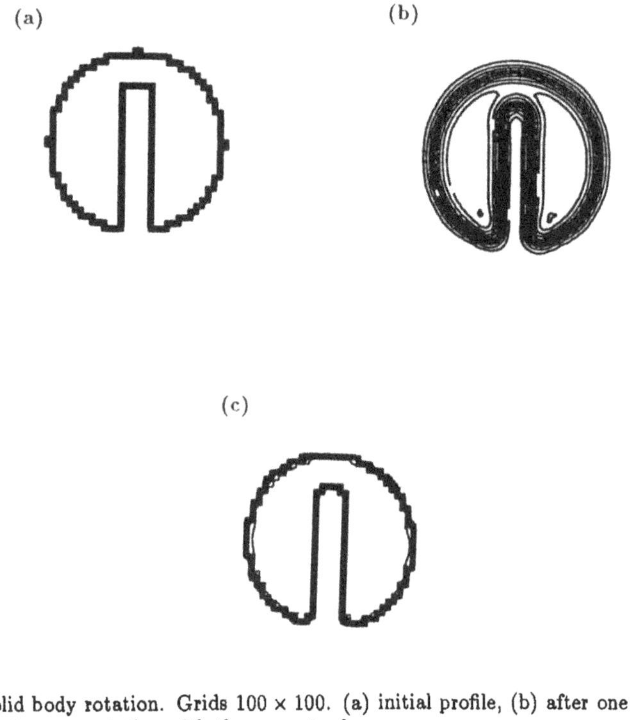

Fig.1 : Solid body rotation. Grids 100 × 100. (a) initial profile, (b) after one rotation with CIP0, (c) after one rotation with the present scheme.

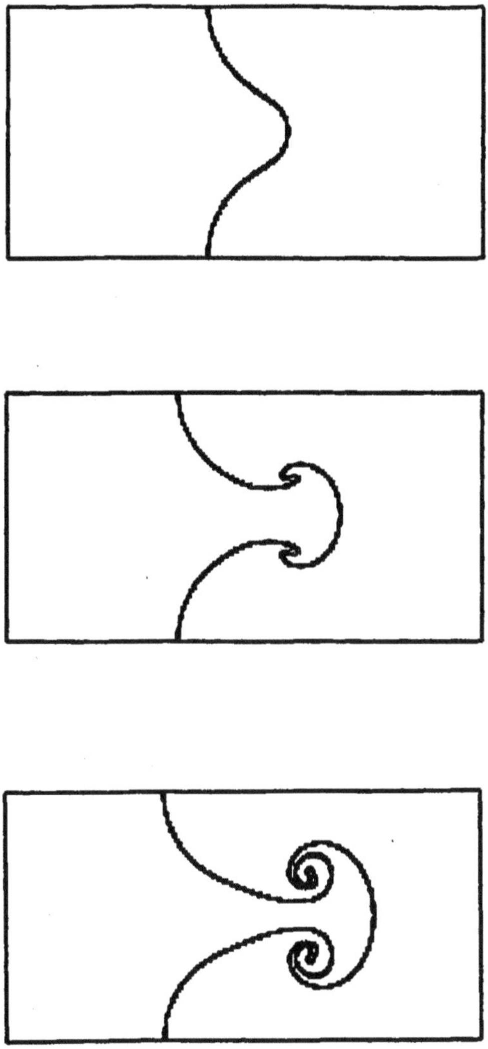

Fig.2 : Density contour for injection of heavier fluid into lighter fluid[7]. Since the process is incompressible, contour lines appear only near the interface. The interface can be described by one grid as shown here. Grids used is 180 × 90.

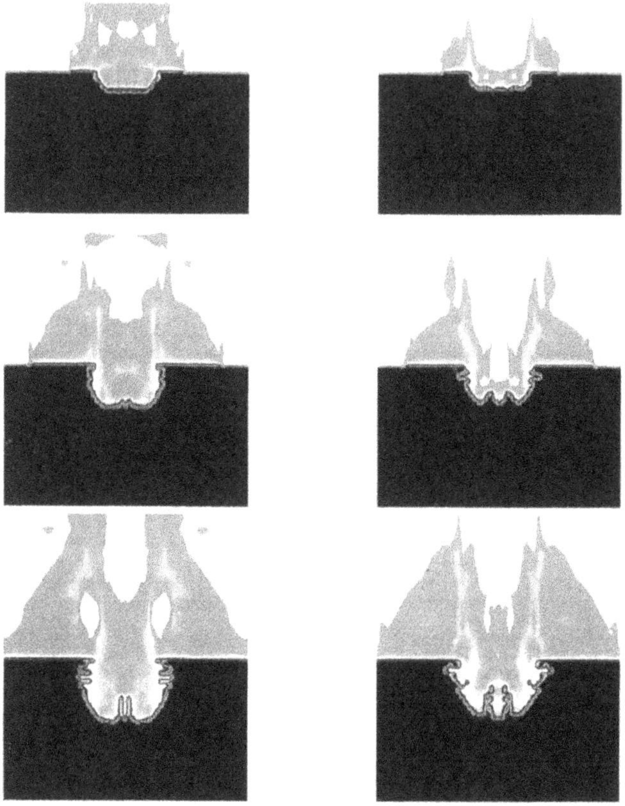

Fig.3 : Isodensity contours at 3.6, 4.8, 6.0 μsec (from top to bottom). Bottom half of a contour is aluminum and laser light is incident from the top. (left) short wavelength laser and (right) long wavelength laser.

ON THE TOTAL VARIATIONAL STABILITY OF NUMERICAL METHODS FOR TWO-DIMENSIONAL CONSERVATION LAWS[1]

ZHAO NING AND WU HUAMO
COMPUTING CENTER, ACADEMIA SINICA, BEIJING, CHINA

1. INTRODUCTION

For the hyperbolic conservation laws in one dimension

$$\frac{\partial u}{\partial t} + \frac{\partial f(u)}{\partial x} = 0 \tag{1.1a}$$

$$u(x,0) = u_0(x) \tag{1.1b}$$

or conservation laws in two dimensions

$$\frac{\partial u}{\partial t} + \frac{\partial f(u)}{\partial x} + \frac{\partial g(u)}{\partial y} = 0 \tag{1.2a}$$

$$u(x,y,0) = u_0(x,y) \tag{1.2b}$$

the numerical methods have been developed rapidly in the recent decade. Especially, for one-dimensional problems, since A. Harten put forward the TVD concept in 1983, the theoretical work about the problems has been founded (see [1]–[5]). However, when the TVD concept was generalized to the multi- dimension case, some essential troubles were yielded. In 1985, J. Goodman and R. LeVeque [6] had proved that conservative TVD schemes in two dimension for (1.2a) have at most first order accuracy. For improving the accuracy of the schemes and perserving the stability of the total variation, Wu Huamo and Yang Shuli [7] put forward the concept of the MmB schemes, B. Cockburn [8–9] constructed the quasi-monotone schemes, and recently F. Coquel and Ph. LeFloch [10] put forward so-called "sharp entropy conditions". Based on the same aim , in this paper, we developed a new kind of stability criteria, so-called the "condition A" and the "condition B". Moreover, we discussed the relationship between these "conditions" and the TVD conditions.

[1]Project supported by The National Natural Science Foundation of China and The State Commission of Science and Technology.

2. One-Dimensional Problems

Let us discuss the numerical methods for the conservation laws (1.1). Partition the space $[0,T] \times (-\infty,\infty)$. Let Δt and Δx be the temporal and spatial step lengths, respectively. Denote the numerical solution at the point (x_j, t^n) by u_j^n. $x_j = j\Delta x, t^n = n\Delta t$ $(j = 0, \pm1, \pm2, ..., n = 0, 1, 2, ...)$, $\lambda = \Delta t/\Delta x$.

Definition 2.1. *For the conservation laws (1.1), a scheme is called conservative , if it can be written as following:*

$$u_j^{n+1} = u_j^n - \lambda(f_{j+\frac{1}{2}} - f_{j-\frac{1}{2}}) \tag{2.1}$$

$$f_{j+\frac{1}{2}} = h(u_{j-k+1}^n, ..., u_{j+k}^n), \ and \ h(u, ..., u) = f(u).$$

Definition 2.2. *A scheme is called TVD, if the following inequality holds for every integer n*

$$\sum_j | u_{j+1}^{n+1} - u_j^{n+1} | \le \sum_j | u_{j+1}^n - u_j^n | \tag{2.2}$$

Definition 2.3. *A scheme is called satisfying the "condition A", if the following inequality holds for every integer n*

$$\sum_j (| u_j^{n+1} - \frac{1}{2}(u_{j+1}^n + u_j^n) | + | u_j^{n+1} - \frac{1}{2}(u_j^n + u_{j-1}^n) |) \le \sum_j | u_{j+1}^n - u_j^n | \tag{2.3}$$

Theorem 2.1. *For conservation laws (1.1), a scheme is TVD if it satisfies the "condition A".*

In general, the conservative difference schemes can be written in the increment form

$$u_j^{n+1} = u_j^n + C_{j+\frac{1}{2}}(u_{j+1}^n - u_j^n) - D_{j-\frac{1}{2}}(u_j^n - u_{j-1}^n) \tag{2.4}$$

It is well-known [1] that the scheme (2.4) is TVD if $C_{j+\frac{1}{2}}$ and $D_{j+\frac{1}{2}}$ satisfy that

$$C_{j+\frac{1}{2}}, D_{j+\frac{1}{2}} \ge 0 \ and \ C_{j+\frac{1}{2}} + D_{j+\frac{1}{2}} \le 1.$$

Theorem 2.2. *If the schemes (2.4) satisfy that*

$$0 \le C_{j+\frac{1}{2}} \ , \ D_{j+\frac{1}{2}} \le \frac{1}{2} \tag{2.5}$$

then, the schemes satisfy the "condition A".

Note, the condition (2.5) is generally correspondent with the condition of TVD schemes, but is more constricted.

3. Two-Dimensional Problems

Partition the space $R^+ \times R^2$. Δt represents the temporal step length, and Δx, Δy the spatial step lengths in x, y-direction, respectively. Denote the numerical solution at the point (x_j, y_k, t^n) as u^n_{jk}. $x_j = j\Delta x$, $y_k = k\Delta y$, $t^n = n\Delta t$ $(j, k = 0, \pm 1, \pm 2, ..., \ n = 0, 1, 2, ...)$ $\lambda = \Delta t/\Delta x, \mu = \Delta t/\Delta y$. So, conservative schemes in two dimensions can be written as

$$u^{n+1}_{jk} = u^n_{jk} - \lambda(f_{j+\frac{1}{2}k} - f_{j-\frac{1}{2}k}) - \mu(g_{jk+\frac{1}{2}} - g_{jk-\frac{1}{2}}) \tag{3.1}$$

where

$$f_{j+\frac{1}{2}k} = h^f(u^n_{j-l+1k-s}, ..., u^n_{j+lk+s})$$

$$g_{jk+\frac{1}{2}} = h^g(u^n_{j-lk-s+1}, ..., u^n_{j+lk+s})$$

and $h^f(u, ..., u) = f(u)$, $h^g(u, ..., u) = g(u)$.

According to the previous TVD concept, we say that a scheme in two dimensions is TVD, if

$$\sum_{j,k}(\Delta y \mid u^{n+1}_{j+1k} - u^{n+1}_{jk} \mid + \Delta x \mid u^{n+1}_{jk+1} - u^{n+1}_{jk} \mid)$$

$$\leq \sum_{j,k}(\Delta y \mid u^n_{j+1k} - u^n_{jk} \mid + \Delta x \mid u^n_{jk+1} - u^n_{jk} \mid) \tag{3.2}$$

Similar to the case in one dimension, the "condition A" in two dimensions requires:

$$\begin{aligned} &\frac{1}{2}(\Delta x + \Delta y)\sum_{j,k}[\mid u^{n+1}_{jk} - \frac{1}{4}(u^n_{jk} + u^n_{j+1k} + u^n_{jk+1} + u^n_{j+1k+1}) \mid \\ &+ \mid u^{n+1}_{jk} - \frac{1}{4}(u^n_{jk} + u^n_{j-1k} + u^n_{jk+1} + u^n_{j-1k+1}) \mid + \mid u^{n+1}_{jk} - \frac{1}{4}(u^n_{jk} \\ &+ u^n_{j+1k} + u^n_{jk-1} + u^n_{j+1k-1}) \mid + \mid u^{n+1}_{jk} - \frac{1}{4}(u^n_{jk} + u^n_{j-1k} \\ &+ u^n_{jk-1} + u^n_{j-1k-1}) \mid] \leq \sum_{j,k}(\Delta y \mid u^n_{jk} - u^n_{j-1k} \mid + \Delta x \mid u^n_{jk} - u^n_{jk-1} \mid) \end{aligned} \tag{3.3}$$

Similar to Theorem 2.1, we have

Theorem 3.1. *If the schemes (3.1) satisfy the "condition A" in two dimensions, then it is TVD in two dimensions.*

Although we can obtain the convergence of numerical schemes under the "condition A" in two dimensions, it is easy to see that the requirement of the "condition A" is too constrict. When the "condition A" in two dimensions degenerates into one dimension, what can be obtained is

$$\sum_j [\mid u_j^{n+1} - \frac{1}{2}(u_{j+1}^n + u_j^n) \mid + \mid u_j^{n+1} - \frac{1}{2}(u_j^n + u_{j-1}^n) \mid]$$

$$\leq \frac{\Delta y}{\Delta x + \Delta y} \sum_j \mid u_{j+1}^n - u_j^n \mid$$

rather than the "condition A" in one dimension. If we take $\Delta x = \Delta y$, it can be obtained that

$$\sum_j [u_j^{n+1} - \frac{1}{2}(u_{j+1}^n - u_j^n) \mid + \mid u_j^{n+1} - \frac{1}{2}(u_j^n + u_{j-1}^n) \mid]$$

$$\leq \frac{1}{2} \sum_j \mid u_{j+1}^n - u_j^n \mid .$$

We consider that the conflict reflects the essential difference of TVD concepts between in one dimension and in two dimensions from one side.

For consistence with the "condition A" in one dimension, we define the "condition B" in two dimension as follow

Definition 3.1. *The schemes (3.1) satisfy the "condition B", if*

$$\sum_{j,k} [u_{jk}^{n+1} - \frac{1}{4}(u_{jk}^n + u_{j+1k}^n + u_{jk+1}^n + u_{j+1k+1}^n) \mid$$

$$+ \mid u_{jk}^{n+1} - \frac{1}{4} u_{jk}^n + u_{j-1k}^n + u_{jk+1}^n + u_{j-1k+1}^n) \mid]$$

$$+ \mid u_{jk}^{n+1} - \frac{1}{4}(u_{jk}^n + u_{jk-1}^n + u_{j+1k}^n + u_{j+1k-1}^n) \mid$$

$$+ \mid u_{jk}^{n+1} - \frac{1}{4}(u_{jk}^n + u_{j-1k}^n + u_{jk-1}^n + u_{j-1k-1}^n) \mid]$$

$$\leq 2 \sum_{j,k} [\mid u_{j+1k}^n - u_{jk}^n \mid + \mid u_{jk+1}^n - u_{jk}^n \mid] \tag{3.4}$$

hold for every integer n.

THEOREM 3.2. *If a scheme in two dimensions satisfies the "condition B", then it also satisfies the "condition A" when it degenerates into one dimension.*

The increment type schemes in two dimensions can be written as follows:

$$u_{jk}^{n+1} = u_{jk}^n + \bar{C}_{j+\frac{1}{2}k}(u_{j+1k}^n - u_{jk}^n) - \bar{D}_{j-\frac{1}{2}k}(u_{jk}^n - u_{j-1k}^n)$$

$$+\bar{C}_{jk+\frac{1}{2}}(u_{jk+1}^n - u_{jk}^n) - \bar{D}_{jk-\frac{1}{2}}(u_{jk}^n - u_{jk-1}^n) \tag{3.5}$$

Thus, we have

THEOREM 3.3. *If the coefficients of the schemes (3.5) satisfy that*

$$0 \leq \bar{C}_{j+\frac{1}{2}k}, \bar{D}_{j+\frac{1}{2}k}, \bar{C}_{jk+\frac{1}{2}}, \bar{D}_{jk+\frac{1}{2}} \leq \frac{1}{4} \tag{3.6}$$

then the schemes satisfy the "condition B".

4. CONCLUSION

In above sections, we have discussed a new kind of nonlinear stability criteria, so-called the "condition A" and the "condition B". In the case of one-dimension, the "condition A" is generally identical with the TVD property. However, in the case of two dimensions, although the "condition A" in two dimensions is associated with the TVD property in two dimensions, it conflicts with the "condition A" in one dimension and high order accuracy. So, we have defined the "condition B" , which is associated with the "condition A" in one dimension and increment type schemes in two dimensions, that is , Theorem 3.2 and Theorem 3.3. However, the convergence of the schemes under the "condition B" in two dimensions still requires further research.

REFERENCES

[1]. A. Harten, High Resolution Schemes for Hyperbolic Conservation Laws. J. Comput. Phys., 49(1983), pp.357-393.

[2]. A. Harten, On a Class of High Resolution Total-Variation Stable Finite Difference Schemes, SIAM J. Numer. Anal. 21,(1984),pp.1-23

[3]. A. Harten and S. Osher, Uniformly High Order Accurte Non-Oscillatory Schemes I, SIAM J. Num. Anal., 24(1987), pp.279-309.

[4]. P.K. Sweby, High Resolution Schemes Using Flux Limiters for Hyperbolic Conservation Laws, SIAM J. Num. Anal., 21(1984), pp.995-1011

[5]. R. Sanders, On Convergence of Monotone Finite Difference Schemes with Variable Spatial Differencing, Math. Comp. 40(1983),pp. 91-105

[6]. J. Goodman and R. LeVeque, On the Accuracy of Stable Schemes for 2D Scalar Conservation Laws, Math. Comp. 45(1985),pp. 15-21

[7]. H.M. Wu and S.L. Yang, MmB - A New Class of Accurate High Resolution Schemes for Conservation Laws in Two Dimensions, Impact Comp. Sci. Eng. vol.1, No.4(1989),pp. 217-259

[8]. B. Cockburn, The Quasi-Monotone Schemes for Scalar Conservation Laws, Part I, SIAM J. Numer. Anal. 26(1989),pp. 1325-1341

[9]. B. Cockburn, The Quasi-Monotone Schemes for Scalar Conservation Laws, Part II, Part III, SIAM J.Numer. Anal. 27(1990),pp. 247-276

[10]. F. Coquel and Ph. LeFloch, Convergence of Finite Difference Schemes for Conservation Laws in Several space Dimensions: The Corrected Antidiffusion Flux Approach, Math. Comp. 57(1990), pp.169-210

www.ingramcontent.com/pod-product-compliance
Lightning Source LLC
LaVergne TN
LVHW082256241026
850861LV00008B/58

* 9 7 8 9 8 1 0 2 1 3 3 9 8 *